T0224277

# Location Systems

*An Introduction to the Technology*
*Behind Location Awareness*

# Synthesis Lectures on Mobile and Pervasive Computing

## Editor

**Mahadev Satyanarayanan,** *Carnegie Mellon University*

Mobile computing and pervasive computing represent major evolutionary steps in distributed systems, a line of research and development that dates back to the mid-1970s. Although many basic principles of distributed system design continue to apply, four key constraints of mobility have forced the development of specialized techniques. These include unpredictable variation in network quality, lowered trust and robustness of mobile elements, limitations on local resources imposed by weight and size constraints, and concern for battery power consumption. Beyond mobile computing lies pervasive (or ubiquitous) computing, whose essence is the creation of environments saturated with computing and communication yet gracefully integrated with human users. A rich collection of topics lies at the intersections of mobile and pervasive computing with many other areas of computer science.

## RFID Explained
Roy Want
2006

## Controlling Energy Demand in Mobile Computing Systems
Carla Schlatter Ellis
2007

## Application Design for Wearable Computing
Dan Siewiorek, Asim Smailagic, and Thad Starner
2008

## Location Systems: An Introduction to the Technology Behind Location Awareness
Anthony LaMarca and Eyal de Lara
2008

Location Systems: An Introduction to the Technology Behind Location Awareness
Anthony LaMarca and Eyal de Lara

ISBN: 978-3-031-01350-8     paperback
ISBN: 978-3-031-02478-8     ebook

DOI 10.1007/978-3-031-02478-8

A Publication in the Springer series
*SYNTHESIS LECTURES ON MOBILE AND PERVASIVE COMPUTING #4*

Lecture #4

Series Editor: Mahadev Satyanarayanan, Carnegie Mellon University

**Series ISSN**
ISSN: ISSN1933-9011  print
ISSN: 1933-902X     electronic

# Location Systems

*An Introduction to the Technology*
*Behind Location Awareness*

**Anthony LaMarca**
Intel Corporation

**Eyal de Lara**
University of Toronto

*SYNTHESIS LECTURES ON MOBILE AND PERVASIVE COMPUTING #4*

## ABSTRACT

Advances in electronic location technology and the coming of age of mobile computing have opened the door for location-aware applications to permeate all aspects of everyday life. Location is at the core of a large number of high-value applications ranging from the life-and-death context of emergency response to serendipitous social meet-ups. For example, the market for GPS products and services alone is expected to grow to US$200 billion by 2015.

Unfortunately, there is no single location technology that is good for every situation and exhibits high accuracy, low cost, and universal coverage. In fact, high accuracy and good coverage seldom coexist, and when they do, it comes at an extreme cost. Instead, the modern localization landscape is a kaleidoscope of location systems based on a multitude of different technologies including satellite, mobile telephony, 802.11, ultrasound, and infrared among others.

This lecture introduces researchers and developers to the most popular technologies and systems for location estimation and the challenges and opportunities that accompany their use. For each technology, we discuss the history of its development, the various systems that are based on it, and their trade-offs and their effects on cost and performance. We also describe technology-independent algorithms that are commonly used to smooth streams of location estimates and improve the accuracy of object tracking. Finally, we provide an overview of the wide variety of application domains where location plays a key role, and discuss opportunities and new technologies on the horizon.

## KEYWORDS

localization, location systems, location tracking, context awareness, navigation, location sensing, tracking, Global Positioning System, GPS, infrared location, ultrasonic location, 802.11 location, cellular location, Bayesian filters, RFID, RSSI, triangulation

# Contents

1. **Introduction** ...................................................................................................... 1
   1.1 Lecture Overview ........................................................................................ 3

2. **The Global Positioning System** ................................................................... 5
   2.1 GPS Origins ................................................................................................ 6
   2.2 System Architecture ................................................................................... 7
       2.2.1 Earth-Orbiting Satellites ................................................................. 7
       2.2.2 Ground Stations .............................................................................. 8
       2.2.3 GPS Receivers ................................................................................ 8
   2.3 Basic GPS-Positioning Algorithm ............................................................ 9
       2.3.1 Satellite Range Estimation ............................................................. 10
       2.3.2 Satellite Coordinate Estimation ..................................................... 11
   2.4 GPS Errors and Biases ............................................................................... 12
       2.4.1 Ranging Errors ............................................................................... 12
       2.4.2 Satellite Geometry Errors ............................................................... 13
   2.5 Differential GPS ........................................................................................ 15
       2.5.1 Real-Time Differential GPS ........................................................... 15
       2.5.2 Real-Time Kinematic ..................................................................... 15
   2.6 Future GPS Enhancements ........................................................................ 16
   2.7 Other Global Navigation Satellite Systems ............................................... 16
   2.8 Summary .................................................................................................... 18

3. **Infrared and Ultrasonic Systems** ................................................................. 19
   3.1 Room-Level Localization via Proximity .................................................... 19
   3.2 Subroom Accuracy With Ultrasonic Time of Flight ................................. 23
   3.3 Absolute Location With Time of Flight and Angle of Arrival ................... 25
   3.4 Comparison of Approaches ........................................................................ 28

4. **Location Estimation With 802.11** ................................................................. 33
   4.1 Signal Strength Fingerprinting ................................................................. 34
   4.2 Signal Strength Modeling .......................................................................... 39

           4.2.1   Constructing the AP database ....................................... 41
    4.3   Privacy Considerations ....................................................... 43
    4.4   Improvements and Variants .................................................. 43

5.  **Cellular-Based Systems** ........................................................... **49**
    5.1   Cell ID-Based ................................................................... 49
    5.2   Radio Modeling Approaches ................................................ 50
    5.3   Assisted GPS ................................................................... 52
    5.4   Signal Strength Fingerprinting ........................................... 54
    5.5   Standardization Efforts and Discussion ................................ 57

6.  **Other Approaches** ................................................................... **59**
    6.1   Instrumented Surfaces ....................................................... 59
    6.2   Vision ........................................................................... 61
           6.2.1   Mobile Cameras ...................................................... 61
           6.2.2   Fixed Cameras ....................................................... 61
           6.2.3   Visual Tags .......................................................... 64
           6.2.4   Practical Considerations .......................................... 65
    6.3   Laser Range Finders ......................................................... 65
    6.4   Audible Sound ................................................................. 66
    6.5   Internet Protocol Measurement ........................................... 66
    6.6   Magnetic Field Strength ..................................................... 67
    6.7   Radio Frequency Identification Tags ..................................... 67
    6.8   Radio From FM to Ultrawide Band ...................................... 69
    6.9   Others Still to Come ......................................................... 72

7.  **Improving Localization Accuracy** ............................................... **75**
    7.1   Smoothing ...................................................................... 75
    7.2   Snapping ........................................................................ 77
    7.3   Fusing and Tracking ......................................................... 78
           7.3.1   Sensor Fusion ........................................................ 80
           7.3.2   Tracking ability ..................................................... 80
           7.3.3   Tolerance to Noise .................................................. 80
           7.3.4   Kalman Filters ....................................................... 81
           7.3.5   Particle Filters ...................................................... 82
    7.4   Summary ........................................................................ 84

**8.    Location-Based Applications and Services** ........................................................ 85
   8.1    Navigation and Way-Finding ............................................................. 85
   8.2    Asset Tracking ................................................................................. 86
   8.3    Emergency Response ......................................................................... 87
   8.4    Geofencing ....................................................................................... 88
   8.5    Location-Based Content and Search ................................................. 90
   8.6    Social Networking ............................................................................ 93
   8.7    Health and Wellness ......................................................................... 94
   8.8    Gaming and Entertainment .............................................................. 96
   8.9    Challenges ........................................................................................ 98

**9.    Challenges and Opportunities** ....................................................................... 101
   9.1    Privacy ............................................................................................ 101
   9.2    "Place" Technologies ...................................................................... 104
   9.3    System Support for Location ........................................................... 105
   9.4    Technologies on the Horizon .......................................................... 106
   9.5    Conclusions .................................................................................... 107

**References** ....................................................................................................... 109

**Author Biography** ........................................................................................... 121

# CHAPTER 1

# Introduction

Fifty years of developments in electronic location technology, coupled with the advent of mobile computing, fueled by smaller, less expensive, and more capable devices, have opened the door for location-aware applications to permeate all aspects of everyday life. Location is at the core of a number of high-value applications including emergency response, navigation, asset tracking, ground surveying, and many others. The market for GPS products and services alone is expected to grow to US$200 billion by 2015 [141]. Location is also commonly used to infer other contexts. For example, symbolic location is often a good proxy for activity (e.g., grocery store is indicative of shopping). Similarly, social roles and interactions can be learned from patterns of colocation [39], and physical activities and modes of transportation can often be inferred from the changes in coordinate-based location [129, 157].

Unfortunately, there is no single location technology that is good for every situation and exhibits high accuracy, low cost, and universal coverage. In fact, high accuracy and good coverage seldom coexist, and when they do, it comes at an extreme cost. The modern localization landscape is thus populated by a kaleidoscope of location systems based on a multitude of different technologies including satellite, mobile telephony, 802.11, ultrasound, and infrared among others.

The aim of this lecture is to inform researchers and developers about the most popular technologies and systems for location estimation, and the challenges and opportunities that accompany their use. Throughout this lecture, we will illustrate the trade-offs made by various systems and describe the effect on cost and performance. For each technology, we introduce the history of its development, the various systems that are based on it, and the performance characteristics of those systems. We characterize each of the technologies we discuss across the following dimensions:

- *Accuracy*: This is the most often-cited metric of location systems and refers to the correctness of a system's location estimates. For coordinate-based location systems, accuracy is expressed as an error distribution, typically in centimeters or meters. It is most common for location accuracy to be expressed as a "median error" which indicates that 50% of the location estimates are at least that accurate. Systems that exhibit normally distributed error often have their error expressed in meters at 1 (or 2) sigma, meaning that 66% (or 95%) of the estimates are at least that accurate. Occasionally, errors for different dimensions (such as horizontal vs. vertical) are called out separately if the system performs significantly different across those dimensions. Because symbolic location estimates (e.g., in the bedroom) are either correct or not, accuracy for symbolic location systems is usually expressed as a percentage. While more accuracy is clearly desirable, the most important factor is whether a system is accurate enough for the application in question. A system with median accuracy of 1 cm, for example, will not generate better driving directions than a system with 1-m accuracy.

- *Coverage*: This refers to the physical area within which the system functions and will produce viable location estimates. This tends to be a descriptive metric, rather than a purely quantitative one; the coverage of GPS, for example, is any place with a clear view of at least four GPS satellites. Most of the systems described in this lecture employ some form of beacon/listener pair. These systems only have coverage when listeners are within range of beacons. Systems that employ beacons with short-range signals (such as ultrasound or magnetic fields) are very hard to deploy with wide-area coverage. Systems with long-range beacons (such as satellites or cell towers), on the other hand, are well-suited to the wide area.

- *Privacy*: One of the key characteristics of a location system is the privacy afforded to the users of the system. In the best case, a user's mobile device can estimate its own location, and neither the infrastructure nor other nearby devices can infer where it is. Fortunately, many popular systems such as GPS and many 802.11-based location systems exhibit this trait. Other systems perform some or all of the location estimation in the infrastructure. While this often comes with benefits in cost, scale, or accuracy, it does put the user in a position of having to trust the system designers and managers.

- *Infrastructure cost*: Related to coverage, infrastructure cost refers to the time and money requirements to deploy and maintain a given location system. Ideally, we could quantify this in dollars per square meter for indoor systems and dollars per square kilometer for wide-area systems. Unfortunately, many of the systems we describe are research prototypes for which pricing is unavailable. For this reason, we simply rate the systems on a four-point scale based on the density and complexity of infrastructure they require. It may surprise the reader to see that we classify GPS as having low infrastructure cost. While the total price tag for GPS is indeed large, when factored over the entire surface of the planet, the deployment cost of $27/km^2$ and yearly maintenance of less than $1/km^2$ make it seem like a bargain [42, 123].

- *Per-client cost*: This metric refers to the incremental cost of adding one more device, person, or object to be located using a system. When special-purpose dongles or active badges are required, the incremental cost per client can be high. Systems that track RFID tags or unmodified cell phones, on the other hand, have extremely low per-client cost.

## 1.1 LECTURE OVERVIEW

This lecture provides a technical introduction to a wide variety of commercial and research systems for performing location estimation. In this section, we have described the dimensions used to characterize the performance of localization systems. The remaining chapters are organized as follows: Chapter 2 describes the widely used NAVSTAR Global Positioning System, better known as GPS. Whereas GPS is highly effective in outdoor areas with a clear view of the sky, the system does not work well indoors and requires additional hardware not yet available on the majority of mobile devices. Chapters 3 through 6 present alternative location technologies designed to address these limitations. Chapter 3 describes how infrared light and ultrasound have been used for indoor location estimation. Chapters 4 and 5 discuss techniques that perform localization by utilizing the existing 802.11 and cellular phone radios that mobile devices use for communication. Chapter 6 closes our discussion of location technologies with a roundup of lesser-used approaches ranging from vision to magnetic field distortion. In Chapter 7, we describe technology-independent algorithms that are commonly used to smooth streams of location estimates and improve the accuracy of object tracking. Chapter 8 provides an overview of the wide variety of application domains where location plays a key role, ranging from the life-and-death context of emergency response to serendipitous social meet-ups. Finally, Chapter 9 closes the lecture with a discussion of the challenges of using location technology, and the opportunities and new technologies on the horizon.

. . . .

# CHAPTER 2

# The Global Positioning System

The NAVSTAR Global Positioning System, or GPS,[1] is by far the most widely used location technology. GPS is used in numerous civilian applications including, shipping, street navigation, recreational boating, ground surveying, and many more. The market for GPS products and services was estimated at 15 billion euros in 2001 and is expected to grow to 140 billion euros by 2015 [141]. It is estimated that by 2020, the number of GPS chipsets will approach three billion [78].

By any measure, GPS constitutes a major engineering undertaking. The U.S. Department of Defense approved the project in 1973, and the system was declared fully operational in 1995. The cost of development of GPS has been reported at $14 billion [123], and its annual operation and maintenance cost is estimated at $500 million [42].

GPS is designed as a passive one-way ranging system where all signals are transmitted by Earth-orbiting satellites, and position determination happens at the receivers. This design makes it possible for GPS to provide worldwide coverage and to scale to an unlimited number of users, while at the same time preserving user privacy [98]. GPS supports a range of location services with accuracies that range from several meters to a few millimeters [48, 162]. Conversely, different types of GPS receivers vary in pricing from a hundred dollars for an inexpensive mass market model to tens of thousands of dollars for the more accurate kinematic solutions used for land surveying.

While GPS is a capable system, it is nevertheless far from perfect. Accurate GPS localization requires an unobstructed view of at least four satellites. GPS signals do not penetrate well through walls, soil, and water, which means that the system cannot be used inside buildings, underground (e.g., inside a mine or tunnel), or for subsurface marine navigation [85]. Signal can also be obstructed by large buildings in the so-called urban canyons created by tall buildings in urban areas. Other localization technologies discussed in subsequent chapters of this lecture are more appropriate for these environments.

---

[1]NAVSTAR (which is not an acronym) was suggested by Mr. John Walsh, a key decision maker in the U.S. Department of Defense. GPS was coined by General Hank Stehling, the Director of Space for U.S. Air Force DCS Research and Development in the early 1970s [127].

In the rest of this chapter, we first provide a brief overview of the origins of GPS. We then describe the GPS system architecture, the basic GPS positioning algorithm, and the biases and errors that affect GPS accuracy. We close with a discussion of the existing and future extensions to basic GPS and a brief introduction to the other satellite-based positioning systems.

## 2.1   GPS ORIGINS

GPS represents the synthesis of three location technologies developed by the U.S. military: *Transit*, *Timation*, and *621B* [52]. Transit, also referred to as the Navy Navigation Satellite System, enabled positioning at sea level within a few hundred meters of precision by measuring the Doppler shift of 400-MHz radio signals broadcast by polar-orbiting satellites. Transit became operational in the 1960s and was still in use until the early 1990s, when it was made obsolete by GPS. Transit, however, had significant shortcomings. Foremost, to avoid radio interference, the system was limited to five simultaneously operating satellites which resulted in limited temporal coverage with windows of unavailability ranging from 35 to 100 min. The two-dimensional nature of the system also kept it from being used for air applications. Nevertheless, Transit contributed significantly to the development of GPS by proving that space system could have excellent reliability, with lifetimes in excess of 15 years.

The Timation system was developed by the Naval Research Laboratory to provide precise time synchronization between points on the Earth. Timation was the first effort to orbit precise clocks, initially using quartz-crystal oscillators, and later shifting to rubidium and cesium atomic clocks. Precise clocks are critical for GPS operation, as they improve the prediction of satellite orbits and enable the GPS satellites to stay tightly synchronized.

Finally, the U.S. Air Force Project 621B introduced a technique for estimating the distance from a satellite by determining the start, or phase, of a repeated pseudorandom noise (PRN) sequence. Properly selected PRN codes have the advantage that they can be decoded even when independent multiple satellites transmit simultaneously on the same frequency. Moreover, 621B showed that PRN codes could be detected even when power density was 1/100th that of the ambient radio noise.

The GPS system design borrowed from these three projects. The signal structure and frequencies were taken from the Air Force's 621B project, satellite orbits are similar to those for the Timation system, and the algorithms for orbit prediction are based on those developed for the Transit system [127].

GPS was designed as a dual-use technology intended for both military and civilian use; however, to maintain an accuracy advantage for the military, the original GPS design included a feature known as Selective Availability, which added intentional noise to the signals to degrade the accuracy of civilian GPS models. Ironically, a shortage of military-grade GPS receivers during the first Persian Gulf War in 1991 forced the U.S. Pentagon to disable Selective Availability for the duration of the war and used civilian receivers to make up the shortfall [123]. Selective Availability was permanently discontinued in May of 2000, and in September 2007, the U.S. Government announced its decision to eliminate the feature from future GPS satellites [160]. In December of 2004, the U.S. government reiterated its commitment to provide GPS access free of direct user fees for civil, commercial, and scientific uses [133].

## 2.2    SYSTEM ARCHITECTURE

The GPS system architecture consists of three distinct parts: a constellation of Earth-orbiting satellites that broadcast a continuous ranging signal, ground stations that update the satellites' coordinate projections and clocks and, finally, the receivers that use the GPS signals to estimate their position.

### 2.2.1    Earth-Orbiting Satellites

The current GPS constellation consists of 31 satellites organized into six non-geostationary circular orbits 26,560 km above the Earth with a 12-h period (see Figure 2.1). Full GPS coverage requires 24 operational GPS satellites. The additional satellites operate as active spares to accommodate occasional maintenance downtime and to assure system robustness [48].

There have been four generations of GPS satellites. The 11 original "Block I" satellites were launched between 1978 and 1985. These were followed by 29 Block II and IIA satellites between

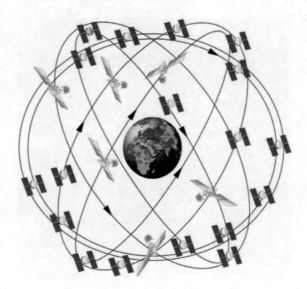

**FIGURE 2.1:** GPS satellite constellation. Source: http://pnt.gov/public/images/. © 1999 IEEE.

1989 and 1997. These original satellites are being gradually replaced by Blocks IIR and IIR-M satellites. Successive generations have had better clocks to increase the length of time they can go between updates and, more recently, additional signals designed to offset the effects of ionospheric and multipath delay (both discussed in Section 2.4). Detailed information about the current GPS constellation is available in the U.S. Naval Observatory Web site at http://tycho.usno.navy.mil/gpscurr.html.

### 2.2.2  Ground Stations

GPS ground stations are responsible for monitoring satellite positions and providing satellites with clock corrections and satellite orbit updates. There currently are enough ground monitoring stations to allow each satellite to be simultaneously tracked by at least two monitoring stations (see Figure 2.2). Simultaneous satellite tracking improves the precision of orbit calculations increasing localization accuracy.

### 2.2.3  GPS Receivers

GPS receivers determine their position by simultaneously tracking at least 4, but commonly up to 12, satellites. GPS receivers can be augmented with other sensors, such as altimeters, accelerometers, and gyroscopes to compensate for gaps in GPS coverage.

**FIGURE 2.2:** GPS ground stations. Shown on the map are the locations of master control station and monitoring stations source: http://www.kowoma.de/en/gps/control_segment.htm.

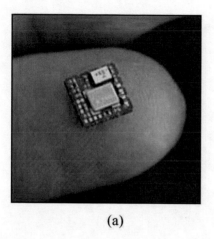

(a)                                                              (b)

**FIGURE 2.3:** Modern GPS receivers. The figure shows (a) an Epson GPS chip and (b) a Globalsat GPS wristwatch.

GPS receivers have benefited greatly from miniaturization. The original mobile units tested by the U.S. Army in the 1970s weighed 25 lbs and filled a backpack [123]. In contrast, modern GPS receivers are typically around the size of a cell phone, and new single chip GPS implementations have made form factors as small as a wristwatch as possible (see Figure 2.3) [97].

## 2.3     BASIC GPS-POSITIONING ALGORITHM

The basic GPS-positioning algorithm is supported by all GPS receivers and allows the receiver to estimate its position in three dimensions (latitude, longitude, and altitude) by tracking four or more satellites. This estimate is computed from the estimated positions of the satellites and the ranges from the receiver to those satellites. In Figure 2.4, $x$, $y$, $z$ represent the location of the receiver, and $x_i$, $y_i$, $z_i$ represent the location of satellite $i$. The satellite locations are learned from the broadcasts from the satellite. $R_i$ is the distance between the receiver and satellite $i$, which is inferred by measuring the transit time of the signal between the satellite and the receiver and multiplying by the speed of light.

Measuring the signal transit time accurately requires the satellite and receiver's clocks to be tightly synchronized. In practice, however, the use of low-cost crystal oscillators in the receiver introduces a bias that makes the distance from the satellite appear shorter or longer than the real value. Fortunately, the receiver-induced bias will be the same across all of the satellites. Satellite-induced clock bias is less of an issue because of their extremely accurate atomic clocks. As a result, the effect of the receiver's clock bias can be removed by treating it as an extra unknown in the location calculation.

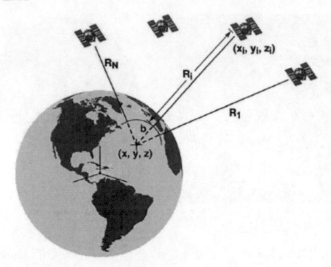

**FIGURE 2.4:** The basic GPS point position algorithm uses the locations of and ranges to the tracked satellites to estimate the receiver's location [42, 113].

$$R_i = \sqrt{(x_i - x^2) + (y_i - y)^2 + (z_i - z)^2} - b \qquad (2.1)$$

The receiver's location in three-dimensional space (x, y, z) and receiver clock bias $b$ is then determined by solving equation (1) for at least four satellites (as opposed to the three satellites that would be required with perfect clocks). In cases where more than four satellites signals are available to the receiver, the redundant data is used to try to identify and eliminate error in the location estimate. This is typically done using a least-squares estimation or with a Kalman filter [100]. Commercial GPS units using this basic algorithm estimate location with median accuracy of around 10 m [153].

## 2.3.1    Satellite Range Estimation

To allow the distance between the satellite and the receiver to be estimated, GPS satellites transmit radio signals modulated by pseudorandom noise (PRN) codes, which consist of binary sequences that appear to be randomly generated. GPS receivers continually compare the signals they are receiving with a locally generated replica of the satellite's PRN code (as illustrated in Figure 2.5). The time delta between the received signal and the local PRN code represents the signal's travel time. Range is computed by simply multiplying the travel time by the speed of light (299,729,458 m/s).

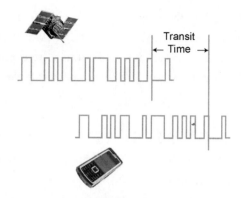

**FIGURE 2.5:** GPS ranging. The figure shows how the receiver can determine the signal transit time by comparing the signal received from the satellite (top) with a replica generated at the receiver (bottom) [42, 113]. © 1999 IEEE.

The use of orthogonal PRN codes allows all GPS satellites to use the same frequencies, while still allowing receivers to differentiate their transmissions and track them in parallel.

GPS satellites each use two PRN codes and modulate them on 1,575 and 1,227 MHz. The civilian PRN code contains 1,023 pulses and repeats every millisecond and is modulated on the 1,575 MHz carrier. A second encrypted PRN code is also sent for military GPS models and is modulated on both frequencies. The military code is a much longer binary sequence composed of $2.35^{10}$ pulses that repeats itself every 38 weeks and is transmitted at a rate that is ten times higher than the civilian code. The higher transmission rate ensures that military GPS users get higher precision than civilian GPS users, and the use of the long encrypted code makes the signal inaccessible to unauthorized users (i.e., receivers that do not know the encryption key cannot differentiate the transmission from thermal noise). Finally, the use of two different frequencies lets military GPS receiver calibrate the ionospheric delay, a key source of ranging error. The theoretical accuracy of GPS ranging is 1/100th of a pulse, or about 3 and 0.3 m for the civilian and military codes, respectively [7]. However, as we will discuss in the next section, GPS ranging accuracy is affected by various biases and errors.

### 2.3.2   Satellite Coordinate Estimation

The GPS receiver obtains the satellite's coordinates from a navigation message which is also modulated on 1,575 and 1,227 MHz signals. The message is encoded within the PRN code at the low bit rate of 50 bits per second. The navigation message is 37,500 bits long and takes 12.5 min to transmit. The message contains all of the relevant information about the GPS constellation: the coordinates of the satellites as a function of time, satellite clock correction parameters, a satellite directory, constellation health status, and parameters for ionospheric error correction. To speed location calculation at the GPS receiver, the satellite coordinates and clock offset are repeated in the navigation message every 30 s.

## 2.4   GPS ERRORS AND BIASES

GPS position accuracy is a function of the error in the ranging estimates and error in the satellite geometry. We discuss each of these in turn.

### 2.4.1   Ranging Errors

Figure 2.6 summarizes the most significant factors that introduce GPS ranging error. As this table shows, ionospheric delay is the dominant source of GPS ranging error. Ionospheric delay, which results from the interaction of the GPS signal with ionized gases in the upper atmosphere, varies based on time of day, time of year, solar flare activity, and the angle of entrance of signal that affects the length of the path through the ionosphere. Figure 2.7 shows the variations in ionospheric delay over a 24-h period. Fortunately, reasonable models of ionospheric delay have been developed, leaving a residual delay with a 4-m mean [110]. Because ionospheric delay varies inversely with the square of the signal frequency, military GPS receivers can factor it out by tracking both the 1,575- and 1,227-MHz signals. This accounts for the majority of the accuracy improvement of military GPS receivers. Tropospheric delay results from the slowing of the GPS signal in the lower atmosphere. Like ionospheric delay, tropospheric delay can be reasonably predicted using mathematical models.

Satellite coordinates errors, which are on the order of 2 m, are the result of the failure of the satellite position models to account for all forces acting on the satellite. Applications that require very

| Error source | One-sigma error [m] |
|---|---|
| Ionosphere | 4.0 |
| Troposphere | 0.7 |
| Satellite position data | 2.1 |
| Satellite clock | 2.1 |
| Multipath | 1.4 |
| **Average Ranging Error** | **5.3** |

**FIGURE 2.6:** The largest sources of ranging errors for civilian GPS [127]. The figure also shows the average ranging error which is the root-sum-square of all sources of ranging errors.

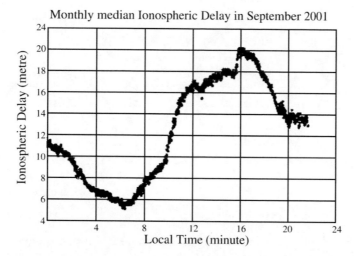

**FIGURE 2.7:** Median of the variation in ionospheric delay over a 24-h period. These measurements were collected in September of 2001 in Thailand's Chiang Mai region [163].

high accuracy, such as geological studies, can eliminate this error almost completely by using precise orbit data (accurate within a few centimeters) that is made available over the Internet [31, 76].

While the satellite's atomic clocks are very stable, they still can accumulate up to 17 ns of error per day, which translates to a range error of 5 m [41]. To correct for this, the satellite clock is continually monitored by the ground monitoring stations, and clock corrections are periodically transmitted.

Multipath error results from reflections of the GPS signal on obstacles, such as buildings and mountains. The magnitude of the multipath error is bounded by transmit time of one pulse, as longer reflections can be detected and ignored. Thus, multipath error can be as high as 1 $\mu$s, which corresponds to 300 m of error. In practice, multipath error is much smaller, adding an average of 1.4 m of error.

Under the assumption that the measurement errors from all satellites are identical and independent, the average ranging error can be calculated as the root-sum-square of the all sources of range error. This quantity is known as the user-equivalent range error (UERE). We next discuss how the UERE, in combination with the satellite geometry, can be used to determine the accuracy of a location estimate.

## 2.4.2   Satellite Geometry Errors

The quality of the GPS location estimation depends on how well the tracked satellites are spread across the sky, and in general, satellite geometry improves as the distance between satellites increases.

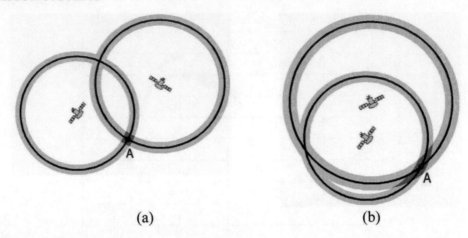

(a)                                        (b)

**FIGURE 2.8:** Effect of good (a) and bad (b) satellite geometry on uncertainty of receiver location. Source: http://www.kowoma.de/en/gps/errors.htm.

Figure 2.8 illustrates the effect of satellite geometry for a simple case with only two satellites. The figure shows the predicted ranges of the two satellites as two bands whose width represents the uncertainty created by the measurement errors. As the figure illustrates, the size of the uncertainty area where the receiver could be located decreases, as the distance between the satellites increases.

The effect of satellite geometry on position accuracy is expressed by a dimensionless metric known as *dilution of precision* or DOP. DOP varies between 1 and 100 based on the satellite geometry, with lower DOP values corresponding to better geometries. DOP is usually expressed as a vertical component (VDOP) and a horizontal component (HDOP). HDOP is always better than VDOP because the satellites are on all sides of the user. In contract, VDOP suffers because GPS receivers can only track satellites that are above the horizon [112].

The accuracy of a GPS position estimate at the one-sigma level can be obtained by multiplying the expected ranging error (UERE) by the appropriate DOP value. Figure 2.9 illustrates the expected position accuracy for sample values of VDOP and HDOP and a UERE calculated based on the errors from Figure 2.6.

|  | One-sigma error |
|---|---|
| User equivalent range error (UERE) | 5.3 m |
| Vertical one-sigma error-VDOP=2.5 | 13.3 m |
| Horizontal one-sigma errors-HDOP=2.0 | 10.6 m |

**FIGURE 2.9:** Estimated GPS position accuracy.

## 2.5    DIFFERENTIAL GPS

Differential GPS takes advantage of the fact that satellite clock and coordinate errors, as well as ionospheric and tropospheric delays, exhibit high temporal and spatial correlation and are very similar even hundreds of kilometers apart [112]. Differential techniques take advantage of this by coordinating multiple GPS receivers that simultaneously track the same satellites. By having one or more GPS receivers fixed at known positions, the observed errors from those receivers can be transmitted to nearby roving GPS receivers. These roving units are then able to reduce their error in proportion to their proximity to the site at which the correction was measured [41].

### 2.5.1    Real-Time Differential GPS

Real-time differential GPS (DGPS) is a relative positioning technique that provides submeter accuracy [112]. In DGPS, a fixed receiver determines the DGPS corrections by comparing its measured satellite ranges with ranges computed using its known coordinates and the satellite coordinates obtained from the navigation message. The DGPS corrections are then transmitted over a ground- or satellite-based wireless link to the rover, which uses them to adjust its ranging measurements. DGPS is offered both as a local and a wide-area service. Maritime DGPS is an example of the former and consists of a network of stations installed at lighthouses along coastal areas in several countries around the world [144]. Each station operates by independently broadcasting real-time DGPS corrections on the 285- to 325-kHz frequency band [172]. These corrections are available at no cost, but require a GPS receiver augmented to receive the corrections.

The Wide Area Augmentation System (WAAS) is a wide-area DGPS implementation that determines GPS range errors at 25 ground base stations. These corrections are broadcast from four geostationary satellites. Measurements from 27 U.S. airports show that WAAS provides accuracy of 1.8 m at least 95% of the time [177]. Other examples of wide area DGPS systems include the European Geostationary Navigation Overlay System (EGNOS), Japan's Multifunctional Transport Satellite Augmentation System (MSAS), and India's GPS and GEO Augmented Navigation (GAGAN).

### 2.5.2    Real-Time Kinematic

Real-time kinematic (RTK) is a GPS variant that achieves subcentimeter accuracy by using an alternative technique for satellite ranging based on measuring the number of fractional and full signal cycles that happen between the satellite and the receiver. The range is then computed by multiplying this value by the carrier's wavelength. While the fraction of the last cycle is easy to determine, the number of full cycles is ambiguous [41]. This ambiguity can be resolved by taking simultaneous measurements at two receivers. Kim and Langley provide a good review of the different ambiguity resolution methods that have been developed [81]. After receiving the base's coordinates and measurements and resolving the ambiguity, roving RTK receivers obtain centimeter-level positioning accuracy. RTK is used in applications that require extreme accuracy such as surveying, aircraft landing, and maritime construction. High-end RTK systems cost tens of thousands of dollars and require line of sight between the coordinating GPS receivers.

## 2.6   FUTURE GPS ENHANCEMENTS

The next generation of GPS satellites will be deployed between 2008 and 2014 and will include new PRN codes that will enable civilian receivers to correct for the ionospheric effects that account for a large portion of the GPS error [56]. The performance of DGPS receivers is currently limited by user equipment and multipath errors, and as such is not expected to be affected by GPS modernization. On the other hand, RTK GPS will benefit from reduced time for ambiguity resolution.

## 2.7   OTHER GLOBAL NAVIGATION SATELLITE SYSTEMS

While GPS is by far the most popular satellite navigation system in use today, other nations have developed, or are in the process of developing, alternative satellite-based positioning system. These efforts are mainly motivated by the reluctance to base critical infrastructure on a service controlled by the armed forces of a foreign nation.

The GLObal'naya NAvigatsionnaya Sputnikovaya Sistema (GLONASS) [53] was originally deployed by the Soviet Union as an answer to GPS and is currently operated by the Russian Federation. While the system fell into neglect after the collapse of the Soviet Union, the Russian Federation, in cooperation with India, has started to rebuild the system. At the time of writing, the GLONASS constellation has 18 satellites, still six short of the required 24 satellites for full worldwide operational capability.

The Galileo system is a 30-satellite location system under development by the European Union. The most remarkable characteristic of Galileo is that it will be under full civilian control under a public–private partnership where the European Commission owns the physical system as a public asset, but a concessionaire is responsible for day-to-day system operation. Galileo's architecture also allows for regional and local augmentation to improve service. Galileo is expected to become operational sometime between 2011 and 2013 [140].

The planned improvements to GPS and GLONASS and the advent of Galileo raise the possibility that, over the next 5 to 10 years, there will be up to 80 satellites available for positioning [141]. It is fair to assume that this will lead to the introduction of multisystem receivers (dual GPS/GLONASS receivers are already available) that take advantage of the additional satellites

| Technology | Basic GPS | GPS + WAAS | Real-time Kinematic GPS |
|---|---|---|---|
| Accuracy | ★★☆☆<br><br>3D coordinates with 10 m median accuracy | ★★★☆<br><br>3D coordinates with 2 m median accuracy | ★★★★<br><br>3D coordinates with 10 cm accuracy |
| Coverage | ★★★☆<br><br>Outdoors with clear view of 4+ GPS satellites | ★★☆☆<br><br>Outdoors in USA — Requires clear view of 4+ GPS satellites and a WAAS satellite | ★☆☆☆<br><br>Outdoors with 4+ GPS satellites and requires line-sight between mobile unit and surveyed unit |
| Infrastructure cost | ★★★★<br><br>$14 B US initial cost + $500 M US yearly for global coverage | ★★★★<br><br>WAAS satellites needed in addition to GPS constellation | ★☆☆☆<br><br>Beyond GPS constellation, requires calibrated, surveyed ground unit |
| Per-client cost | ★★★☆<br><br>GPS antenna and chipset required | ★★★☆<br><br>GPS antenna and WAAS-capable chipset required | ★☆☆☆<br><br>Special RTK unit required |
| Privacy | ★★★★<br><br>Location is estimated passively on the GPS unit | ★★★★<br><br>Location is estimated passively on the GPS unit | ★★★★<br><br>Location is estimated passively on the GPS unit |
| Well-matched use cases | Outdoor navigation for land, sea and air, emergency response, turn-by-turn driving directions, outdoor mapping/information/tour guide services, personell/pet tracking, fitness/activity tracking, gaming | Outdoor navigation for land, sea and air, emergency response, turn-by-turn driving directions, outdoor mapping/information/tour guide services, personnel/pet tracking, fitness/activity tracking, gaming | Outdoor navigation requiring extreme accuracy, surveying, aircraft landing, maritime construction |

FIGURE 2.10: A summary of the performance characteristics of some common GPS variants.

to increase positioning accuracy, availability, and robustness. Rizos et al. [141] estimate that, by combining GPS, GLONASS, and Galileo, a ground-based receiver should always be able to see 21 satellites except at extreme latitudes compared to just 6 for the GPS-only solution. Similarly, Feng estimates that a combined GPS and Galileo system should have a median error of approximately 1.5 m [43].

## 2.8   SUMMARY

In this chapter and those that follow, we summarize the performance characteristics of the technologies we introduce in a table. In these tables, we rate technologies on a four-point scale from best to worst. For accuracy, "best" indicates the lowest error rates; for coverage, "best" indicates function over a wider range of situations; for cost, "best" indicates the lowest money and effort investment. Figure 2.10 shows the performance summary for a number of GPS variants.

•    •    •    •

CHAPTER 3

# Infrared and Ultrasonic Systems

In this chapter, we describe the various systems that use ultrasound and infrared light to perform localization. This may seem like an odd pairing, as light and sound have very different properties. What they have in common, however, is that while they move freely in open space, light and sound are both largely blocked by the materials such as walls, curtains, and partitions that humans use to define the borders of their living spaces. This makes these two types of signals well suited for indoor location systems as they can determine which room a user is in with high accuracy. It is worth emphasizing the importance of room-level location information for many context-aware applications and services. While an application like navigation requires absolute locations, many others rely instead on the semantics associated with a room name or room type. A location system with a median accuracy of, say, 0.5 m may seem more desirable than a system which is only able to determine which room a user is in; however, you may rethink this assessment once you consider that the former system may incorrectly infer that the user is in the adjoining kitchen 25% of the time, when infact they are in the hallway.

The systems we present in this chapter all use infrared or ultrasound beacons and listeners to determine a location. In some cases, the beacons are embedded in the environment, and the listeners are mobile, and in other cases, the reverse is true. Some systems use radio signals to coordinate the location estimation, while others rely exclusively on the ultrasound or infrared signals. Some systems provide symbolic room-level accuracy. Others use time-of-arrival or angle-of-arrival to determine the user's location within the room with 5–10 cm accuracy. In this chapter, we describe these systems and the approaches they employ, and we discuss how their designs affect the accuracy, coverage, privacy, and scalability of the resulting systems.

## 3.1 ROOM-LEVEL LOCALIZATION VIA PROXIMITY

The first indoor location system, the Active Badge system, was developed in 1990 at the Olivetti Research Laboratory [172]. The Active Badge system determines the room in a building where people are located by having each person wear a small (40-g) badge-like device (shown on the right in Figure 3.1) that sends a unique pulse-width modulated infrared code every 15 s. These codes are received by base stations (shown on the left in Figure 3.1) mounted on the ceilings of offices,

**FIGURE 3.1:** An Olivetti Active Badge (right) and the sensor base station (left) used to watch for badges [68].

hallways, and meeting rooms. These base stations decode the received infrared signals and forward the decoded badge-IDs to a centralized location server using the building's wired networking infrastructure. Because base stations are mounted in known locations and because the signals rarely propagate beyond a single room, the server can infer with high probability in which room the badge is beaconing. Because the infrared badges take only 0.1 s to transmit their unique code, collisions between simultaneously beaconing badges are unlikely. In the unlikely event of a collision, the system relies on the clock skew between badges to resolve the problem.

Because infrared signals are easily blocked by fabric, the badges need to be worn on the outside of people's clothing. It was also observed that badges clipped to belts or pockets were often obscured by tables and desks when people were seated. As a result, "clipped to a breast pocket" emerged as the best position in which to wear the Active Badge. The simple design and use of low-power parts enable an Active Badges to run for more than a year before requiring the batteries to be changed. Due to the simplicity in design and architecture, commercial infrared badge systems as those sold by Versus [4] (show in Figure 3.2) have changed little in the nearly 20 years since the Active Badge system was developed.

One disadvantage of the Active Badge system is that it requires specialized infrastructure in the form of base stations to listen for badge IDs. It also requires users to purchase and carry the badge itself. Madhavapeddy et al. explore the possibility of building a room-level accurate location system using ultrasonic signals sent and received from people's existing computing devices [106]. In their system, ordinary PCs repeatedly play an audio track containing an ultrasonic encoding of the room's ID. User's laptops and PDAs listen for and decode these IDs to determine which room they are currently in. Madhavapeddy et al. observed that while the audio capabilities in PCs, laptops, and PDAs are designed to send and receive signals in the human-audible range, they also extend into the ultrasonic range. Their experiments showed that computer speakers and microphones could reliably

**FIGURE 3.2:** A commercial infrared badge, the Versus Personnel Alert Badge is marketed to hospitals to track employees and equipment. Source: http://ascc-inc.com/HealthCare/Versus.htm.

send and receive a 21.2-kHz audio signal which could not be heard by adults. They also confirmed, as can be seen in Figure 3.3, that these ultrasonic signals do not extend beyond the room in which they are played. Reliably transmitting digital data via ultrasound is notoriously tricky because of the large amount of interference created by everyday objects such as jangling keys [106]. As a result, the system of Madhavapeddy et al. was only able to reliably send 8 BITS per second at a distance of 3 m. Despite this seemingly glacial data rate, their system is able to send a unique room ID in 3.5 s.

The WALRUS system is the last proximity system we present, and like the system of Madhavapeddy et al., it uses PC and PDA speakers and microphones to send and receive ultrasonic signals to convey proximity. Unlike the other systems in this subsection, WALRUS does not encode the room's ID using the ultrasonic or infrared beacon signal [16] but, instead, encodes it using a

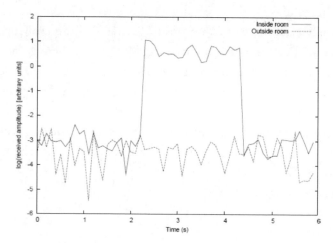

**FIGURE 3.3:** This graph shows the received strength inside and outside a room of a 2-s burst of 21-kHz ultrasound from a speaker in the room 3 m from the wall [106].

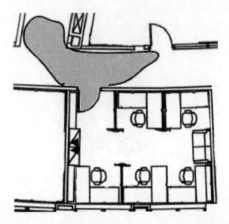

**FIGURE 3.4:** The shaded area in the floorplan shows the extent to which WALRUS's ultrasound beacons extend beyond the room with the door open [16]. © 2005, ACM, Inc. Reprinted by permission.

radio side channel. In the WALRUS system, the PCs that are serving as beacons periodically emit a 10-ms pulse of 21-kHz ultrasound. At the same time, they send an 802.11 datagram containing the rooms' identifying information.[1] As the ultrasounds and radio datagram are transmitted at the same time, WALRUS clients do not need to constantly monitor their microphones. Rather, they can listen for an 802.11 datagram and, only when received, listen for an ultrasound signal. In addition, as no actual data is encoded in the ultrasound, the client's audio task is reduced to simply looking for sufficient energy in the 21-kHz band to connote proximity. This also extended the useful range of the ultrasound signals generated by the beacons to 13 m in their experiments.

Because beaconing PCs are not coordinated in WALRUS, there is the potential for client devices to receive the radio datagram from one beacon and associate it with the ultrasound from a different beacon, in effect, incorrectly identifying which room it is in. Their experiments show that while this is unlikely when only a small number of WALRUS beacons are within 802.11 range of each other, by the time the number of beacons grows to 25, the error rate increases to more than 50% [16]. This issue can be addressed by using a central coordinator to schedule the PC beacons. In the next section, we will also describe a clever technique used by the Cricket system to eliminate these errors and still remain decentralized.

---

[1] In the event that a WALRUS server does not have a wireless interface, it can request that a nearby PC or access point transmit the datagram in its stead.

Borriello et al. measured the extent to which WALRUS's beacons extended beyond room boundaries. Figure 3.4 shows that even with the door open, the ultrasound beacons do not extend much beyond the immediate vicinity of the doorway.

## 3.2   SUBROOM ACCURACY WITH ULTRASONIC TIME OF FLIGHT

The Cricket system, developed as part of the MIT Oxygen project, sought to enable subroom-level location using ultrasound and to do it in a privacy preserving, decentralized fashion [136]. In the Cricket system, beacons are placed on ceilings or high on the walls in the environment and advertise their identity using a combination of RF and ultrasound. Like WALRUS beacons, Cricket beacons encode their identity using RF and send ultrasound pulses containing no digital data. Unlike the systems we have described thus far, the Cricket designers wanted to be able to place multiple Cricket beacons in a single room to allow Cricket listeners (shown in Figure 3.5) to figure out which part of the room (which seat at the table, which work area, which bed, etc.) the user was occupying. To do this, the Cricket system uses of the fact that relative to modern electronics, ultrasound travels very slowly. (In room-temperature air, sound travels at only 0.34 m/ms.) On receiving the radio ID from a beacon, Cricket listeners not only listen for an ultrasound pulse, but time how long it takes to arrive. Because the radio signal travels almost instantaneously relative to the ultrasound, the delay between start of the radio transmission and the observation of the ultrasound pulse is a close approximation of the time of flight of the ultrasound which can, in turn, be converted into an estimate of the distance the ultrasound traveled. By estimating and comparing the distances of the beacons being observed, a Cricket listener can accurately predict which beacon is the closest. Thus, Crickets can be used as a room-level location system, when beacons are sparsely deployed, and as a subroom-

**FIGURE 3.5:** MIT Cricket implementations [136].

level system, when multiple beacons are placed in a room. Subsequent extensions to Cricket show that with sufficient beacon density, their system could determine location within several centimeters and orientation to within 5° [135].

One of the major concerns with using time of flight as a distance metric is *multipath*, or the phenomenon that a signal may be received multiple times due to reflections of the signal in the environment. The Cricket system mitigates errors due to multipath in two ways. First, the Cricket designers observed that the ultrasound signals would commonly bend around the edge of obstacles due to diffraction. As a result, they found it was uncommon to receive a pulse via an indirect path without first receiving it from the direct path. Thus, the Crickets could simply measure the time to the first ultrasonic pulse and ignore the subsequent reflections. Second, Crickets perform smoothing by estimating the closest beacon to be the one with the lowest average estimated distance over a window of samples. This smoothing allows the system to tolerate the occasional ranging measurement error. In practice, these simple heuristics worked well: with a window size of 20 samples, a listener 2 m from a pair of beacons 1.3 m apart was able to able to distinguish which beacon was closer with better than 99% accuracy [136].

One of the most clever aspects of the Cricket system is the way it prevents listeners from associating the infrared pulse from one beacon with the radio broadcast from another. Given that the maximum effective ultrasonic range of the Crickets is 10 m, the Crickets beacons intentionally

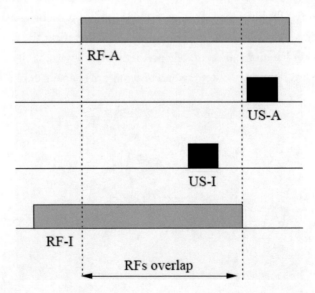

**FIGURE 3.6:** The Cricket system fully overlaps each source's radio transmissions (RF) with the source's ultrasonic pulses (US). This ensures that any ultrasonic pulses in flight at the same time are ignored, as the radio collision can be detected [136]. © 2000, ACM, Inc. Reprinted by permission.

transmit their ID using a slow radio encoding that takes as long as ultrasound takes to propagate 10 m. This ensures that the radio transmission from each cricket beacon fully overlaps the flight of the ultrasound pulse from the same source (illustrated in Figure 3.6). This overlap guarantees that at the time a listener receives an ultrasonic pulse from a beacon, it will still be receiving that beacon's ID via radio. This allows the listeners to detect (and ignore) the RF collision that would occur if two or more ultrasound packets were in flight at the same time. This ensures that pulses will never be incorrectly associated with the wrong beacon ID. To reduce the chance that these collisions occur, Cricket beacons randomize the time between transmissions. (In their experiments, beacons waited between 150 and 350 ms between broadcasts.)

## 3.3    ABSOLUTE LOCATION WITH TIME OF FLIGHT AND ANGLE OF ARRIVAL

Thus far, the systems presented in this chapter estimate the room or portion of a room where the device is located. The final two systems we present estimate with high accuracy the absolute location of a device within a room (e.g., "110 cm south, 50 cm east, and 35 cm up from the northwest corner of Conference Room 206"). The first is the Active Bat system, and it was developed by AT&T Research in 1999. The Active Bat system uses ultrasound receivers to localize small (35 g) pager-like devices called *Bats* (shown in Figure 3.7) in three-dimensional space [66]. The system is extremely accurate, estimating location correctly within 5 cm 50% of the time and within 9 cm 95% of the time. Like Crickets, Bats are located by measuring the travel time of an ultrasonic pulse. In contrast to Crickets, however, the Bats emit the ultrasonic pulses, and receivers in the infrastructure listen for and time the pulse's travel.

**FIGURE 3.7:** An AT&T Active Bat [66]. © 2002, ACM, Inc. Reprinted by permission.

The accuracy of the Active Bat system derives from its dense deployment of ultrasound receiver units (shown in Figure 3.8). In the original AT&T deployment, 100 ceiling-mounted receivers were used in a 100-m² office space. This dense receiver deployment makes it likely that a Bat's pulse would be heard by at least three and likely many more receivers. To estimate location, receivers in the Active bat systems pool their time-of-flight observations to a central server that performs *multilateration* (a generalization of trilateration) in which location is estimated from three or more ranging estimates from known locations [66]. As in the Cricket system, reflections of ultrasound are common, and the Bat system uses the redundancy of the grid of receivers to reject these reflections as outliers. The success of their approach can be seen in the error CDF shown in Figure 3.9.

The Active Bat system is sufficiently accurate that it is possible to place multiple Bats on a rigid object and, by measuring the Bat's location, reconstruct not only the object's location but also its orientation. Experiments in which two Bats were attached to a rigid object 22 cm apart showed that orientation could be estimated within 10° over 80% of the time [66].

The accuracy of the Bat system is dependent not only on the density of ultrasound receivers but also on centimeter-accurate knowledge of each receiver's location. For the University of Cambridge's Bat deployment, this was done with the aid of professional surveying equipment. However, Scott and Hazas showed that it is possible for systems like an Active Bat deployment to be self-surveying [151]. They placed a 1 × 1-m wooden frame fit with 21 Bats in a known geometry in a room and collected 20 min of raw time-of-flight measurements between the Bats and the receivers. By

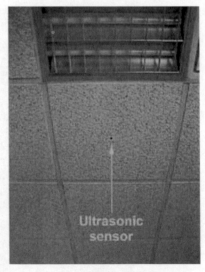

**FIGURE 3.8:** Ultrasound receiver unit for Active Bats. Shown sitting on ceiling tile (left) and shown when installed (right) with only receive sensor exposed [66]. © 2002, ACM, Inc. Reprinted by permission.

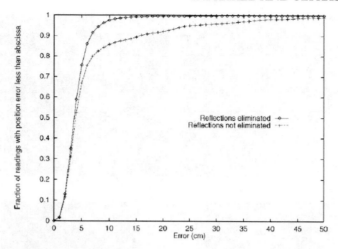

**FIGURE 3.9:** The cumulative distribution function of localization error for the Active Bat system [66]. © 2002, ACM, Inc. Reprinted by permission.

running the Active Bat system's nonlinear regression algorithms over these readings, they were able to infer the ultrasonic beacon locations with accuracy comparable to the manual survey.

The Active Bat system is centrally coordinated, and Bats wishing to be tracked register their interest with a central service via radio. This service employs a slotted schedule to ensure that, at most, one Bat at a time has an ultrasonic pulse in flight. The system designers determined that ultrasonic reverb can take up to 20 ms to die off in an indoor environment; thus, they use a schedule with 50 slots per second. At the start of a slot, the central server requests (via radio) that the Bat with a given ID send an ultrasound pulse. The server then records the elapsed time and ID of each receiver that reports having heard the pulse. At the end of the slot, the server can compute the distance estimates and, using the receiver's locations, can estimate the Bat's location. The Bat system does not need to employ a uniform schedule, and Bats assigned to people can be localized with a higher frequency than those attached to less mobile objects like printers.

The last system we introduce in this chapter is ALTAIR, an indoor location system that augments laptops and PDAs with infrared transmitters and uses a simple image processing algorithm to determine their location [147]. Like the Active Bat system, ALTAIR is centrally coordinated and also uses a slotted "roll call" schedule to have each device in the system blink its infrared LED in turn. Because they are localizing computing devices, ALTAIR uses the computer's wireless networking to signal the clients to blink and uses the DHCP server's client table to know which devices should be polled.

ALTAIR uses wide-angle CCD video cameras fitted with IR-filters to observe the blinking IR-LEDs. These cameras are mounted in the upper corners of rooms pointing in to minimize the chance of the blinking lights being occluded. A sample image from an ALTAIR camera is shown,

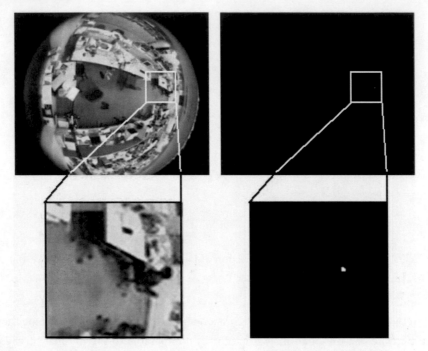

**FIGURE 3.10:** Sample images from an ALTAIR camera before IR filtering (left) and after (right). The blinking IR-LED is clear in the zoomed filtered image [147]. © 2003 IEEE.

both with and without IR filtering in Figure 3.10. In the filtered image on the right, the location of the IR-LED is easy to identify. By watching the video for blinking pixels on the filtered video, the system can infer from the camera's location, which room the polled device is in. In addition, the wide-angle lenses used in ALTAIR have the property that the angle of the infrared light's arrival at the camera can be estimated from the distance of the lit pixel from the center of the image. This allows, with observation from two or more cameras, to estimate the 3-D location of the mobile device. With an observation from only one camera, the 2-D location of the mobile device can be estimated with an assumption of the device's height from the floor. In experiments with two cameras in a 6 × 6-m laboratory, ALTAIR achieved an average accuracy of 9 cm performing 3-D location estimation. Like the Active Bat system, the accurate measurement of the location (and orientation) of the ALTAIR cameras is essential to good system performance. Chapter 6 discusses more examples of systems that use machine vision to determine location.

## 3.4   COMPARISON OF APPROACHES

Three of the systems presented in this chapter, WALRUS, the system of Madhavapeddy et al., and Crickets are completely decentralized and make no use of networking or infrastructure beyond that in the beacons and listeners. That makes these systems suitable for small, grassroots deployments without requiring organizational coordination. The Active Badge system makes use of a central server to collect badge sightings and network infrastructure to connect the servers and infrared receivers. That said, multiple Active Badge deployments could peacefully coexist in the same space if badge ID conflicts were avoided. The most accurate systems turn out to be the most centralized: the ALTAIR system and the Active Bat system rely on a single, coordinated deployment to function properly.

Interestingly, there is no strong mapping between degree of centralization and scalability. While completely decentralized, the WALRUS system drops in accuracy when scaled beyond a few nearby rooms due to the lack of beacon coordination or interference recognition. The most centralized systems employ a slotted schedule to avoid conflicts. This actually ensures that the system will scale gracefully with additional clients coming at the cost of a slightly lower refresh rate for everyone. (We should point out that while the Bat system employs a schedule with 50 slots per second, the ALTAIR prototype used a schedule with only 2 slots per second.) The Active Badge system should be quite scalable: because badges transmit less than 1% of the time, the system should be able to accommodate tens of badges per room with tolerable interference, and the system incurs no interference between badges in different rooms. Both Crickets and the system of Madhavapeddy et al. should scale extremely well, with no interference between the listeners to be located.

With respect to privacy, WALRUS, Crickets, and the system of Madhavapeddy et al. were all designed to allow clients to passively monitor the environment and estimate their own locations; thus, these systems preserve the user's privacy. The three other systems (Active Badge, Active Bat, and ALTAIR) all employ a central server to compute and disseminate the users' locations making these low in privacy. That is not to say that secure, trustworthy systems could not be built on top of these location systems. It is the case, however, that users of these systems have to rely on the good intentions and competence of the system managers to preserve their privacy.

| System | Location Technology | Sender of ID (IR: Infrared US: ultrasound) | Inference avoidance / detection | Location computed by |
|---|---|---|---|---|
| Active Badge | Infrared | badge ID by IR | Collision detection | Server |
| Madhavapeddy et al. | Ultrasound | beacon ID by US | Collision detection | Client |
| WALRUS | Ultrasound+RF | beacon ID by RF | None | Client |
| Cricket | Ultrasound+RF | beacon ID by RF | Collision detection | Client |
| Active Bat | Ultrasound+RF | Bat ID by RF | Central schedule | Server |
| ALTAIR | Infrared+RF | Device ID by RF | Central schedule | Server |

**FIGURE 3.11:** A summary of the designs of the infrared and ultrasonic systems described in this chapter.

In terms of coverage, while all of these systems could conceivably be scaled to a campus or even a city's worth of buildings, they are really only suited for indoor localization. WALRUS and the system of Madhavapeddy et al. are the two that exclusively reuse hardware found in existing devices, making them software-only deployments. This makes these the two easiest systems to deploy

| Technology | Infrared proximity (e.g., Active Badge) | Ultrasound proximity (e.g., WALRUS) | Ultrasound TOF (e.g., Active Bat) | Infrared triangulation (e.g., ALTAIR) |
|---|---|---|---|---|
| Accuracy | ★★★☆ Room ID with high accuracy | ★★★☆ Room ID with high accuracy | ★★★★ 3D location with 5 cm accuracy | ★★★★ 3D location with 9 cm accuracy |
| Coverage | ★★☆☆ Indoor only in room fit with IR receiver/beacon | ★★☆☆ Indoor only in room fit with ultrasonic beacons | ★★☆☆ Indoor only in room fit with ultrasonic infrastructure | ★★☆☆ Indoor only in room fit with infrared cameras |
| Infrastructure cost | ★★☆☆ Infrared receiver or beacon required for each room | ★★☆☆ 1 or more ultrasonic receiver or beacon required for each room | ★☆☆☆ Requires dense array of ultrasonic receivers | ★☆☆☆ Requires 2+ calibrated IR cameras per room |
| Per-client cost | ★★★☆ Inexpensive IR badge/dongle required | ★★★★ Software only solution on device with microphone | ★★★☆ Inexpensive ultrasonic badge/dongle required | ★★★☆ Inexpensive IR badge/dongle required |
| Privacy | ★★★★ If localization is performed on the client. Otherwise ★★☆☆ Opt-out easy by removing badge | ★★★★ Localization is performed by mobile client | ★★☆☆ Localization is performed by infrastructure. Opt-out easy by removing badge | ★★☆☆ Localization is performed by infrastructure. Opt-out easy by removing badge/dongle |
| Well-matched use cases | Asset and personnel tracking, indoor mapping/ navigation/tour guides | Asset and personnel tracking, indoor mapping/ navigation/tour guides | Asset and personnel tracking, tangible UIs, fine-grained info services | Asset and personnel tracking, tangible UIs, fine-grained info services |

**FIGURE 3.12:** A summary of the performance characteristics of the infrared and ultrasonic location estimation techniques.

on a large scale. The other four systems all require custom hardware in the form of badges, cameras, and base stations. Two of these systems, ALTAIR and Active Bats, also require a significant calibration process. In this respect, the accuracy of the systems in this chapter are well correlated with their deployment cost ranging from easy to deploy room-level systems to expensive, carefully tuned centimeter-accurate systems. We close this chapter with tables summarizing some of the key design decisions (shown in Figure 3.11) and the performance characteristics (shown in Figure 3.12) of the systems we have covered.

•　•　•　•

CHAPTER 4

# Location Estimation With 802.11

Estimating location with 802.11 was first proposed in 2000 by Bahl et al. [8], and since then, many variants have been developed. At their most basic, all of these systems work the same way: 802.11 radios and their supporting drivers allow a device to scan for nearby 802.11 access points (APs). Regardless of the variant of 802.11 (a, b, g, n) or whether encryption is enabled, these scans return a list of the nearby APs and their unique IDs called MAC addresses. All of the 802.11 location systems capitalize on the fact that 802.11 access points have limited range (typically less than 100 m), and if a device can hear an access point, it knows that it is in the vicinity of that access point. For these systems, a lighthouse metaphor works well: if you can see the lighthouse, you know you're getting close to shore. Of course, if a device can see more than one AP, it can estimate its own location more precisely, and additional information, such as received signal strength and packet loss rate, can be used to improve the accuracy of location estimates further still. The appeal of using 802.11 for location estimation is strong: nearly all PDAs and notebook computers have built-in 802.11 and increasingly smaller devices like cell phones and mp3 players are also incorporating 802.11. In concert with client availability, 802.11 networking coverage has grown near ubiquitous in densely populated areas; a recent paper cited over a half-million known mapped 802.11 access points in the Tokyo metropolitan area [139]. Densities of this kind raise the possibility of high-coverage indoor/outdoor location with no additional location infrastructure. This near-pervasive client hardware and infrastructure for doing 802.11-based location is the reason why dozens of such location systems have been developed.

To allow client devices to discover and associate with an access point, the 802.11 protocol includes beacon frames that APs can send to alert clients to their presence. The frequency with which these beacon frames are sent varies by model of AP and commonly ranges from tens to hundreds of beacon frames per second. These frames contain the human readable SSID of the network "Joe's hotpot" and the MAC address of the AP and whether the AP is running encryption or not. Client devices can passively learn about nearby APs by listening for a small window of time on each of the 802.11 channels. Alternatively, clients also have the option of initiating an active scan by sending a probe request which prompts access points to reply with a probe response very similar to a beacon frame. Figure 4.1 shows an example of the results of a passive or active scan by an 802.11 client. We

| SSID | MAC | RSSI | Encrypted? |
|------|-----|------|-----------|
| Capital City WiFi | 00:0f:34:ab:0c:e0 | -66 dB | No |
| First Security Trust | 00:0f:f7:0c:e9:c0 | -78 dB | Yes |
| First Security Trust | 00:0f:f7:0c:4f:03 | -105 dB | Yes |

**FIGURE 4.1:** An example of what is returned from an 802.11 scan by a client device. The table shows the service set identifier (SSID) of the wireless network to which the access point belongs, the MAC address of the access point, the received signal strength indication (RSSI) of the power level at which the beacon was received at the client, and whether encryption is enabled.

now delve into the different techniques that have been used to estimate location from the results of such scans.

## 4.1  SIGNAL STRENGTH FINGERPRINTING

The first 802.11 location system, RADAR, was developed by Bahl et al. [8] at Microsoft Research. RADAR and the other systems like it are the most accurate 802.11-based location techniques and are capable of tracking device with 1- to 3-m accuracy depending on the environment [93]. RADAR takes advantage of two properties of the 802.11 signals observed by client devices: spatial variability and temporal consistency. First, due to the short range of 802.11 and the way signals are blocked and reflected by obstructions in the environment, the strength with which an AP is observed varies considerably spatially, even over distances as small as a meter. We have all observed this when we found we could not get good network connection in one place, and we moved over one seat, and the connection improved drastically. Second, the strength of an 802.11 signal tends to be temporally consistent. If a given seat at a table is a good place to establish a connection to a particular AP today, it will likely be a good place to connect in a few minutes, the next day, or the next month. This effect is shown in Figure 4.2 in which a stationary client consistently observes a relatively strong and stable (within 5 db) signal from a specific AP. Using both of these properties, Bahl et al. observed that it is possible to build a map of the radio environment by collecting location-tagged "radio fingerprints" and later use this map to perform localization of mobile 802.11 devices.

Location estimation using radio fingerprints involves two phases: the mapping phase and the location estimation phase. In the mapping phase, a site survey is performed, in which the visible APs and their observed signal strengths are recorded along with the location in which the observation was taken. To provide good coverage, the radio map readings typically span the entire physical space. To provide good accuracy, the readings need to be of sufficiently high density; typically, a reading is collected every few square meters. Figure 4.3 shows a map of an indoor space and a reasonable grid of places to take radio map readings. Given this radio map, estimating location is straightforward: a client in an unknown location performs a radio scan and estimates its location

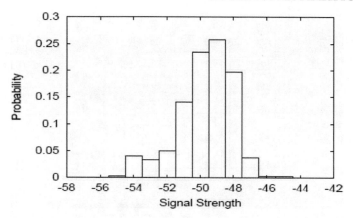

**FIGURE 4.2:** The signal strength observed by a stationary client from a single 802.11 AP over a series of readings is stable within ±5 db [176]. © 2003 IEEE.

to be the place on the radio map whose scan most closely matches its observation. Spatial variation ensures that there will be enough variety in the radio environment to distinguish places from each other, and the temporal consistency means these distinctions are likely to not change between the time of mapping and the location estimation.[1]

RADAR and most of the follow-on fingerprinting systems use the Euclidian distance in signal space as a measure of the similarity of the two radio scans. For example, the distance in signal space for the pair of scans shown in Figure 4.4 is $\sqrt{5 \cdot 5 + 2 \cdot 2 + 10 \cdot 10} = 11.4$.

The simplest algorithm for estimating location called nearest neighbor is not much more complicated than this example: the device's location is estimated to be the location of the mapping scan with the smallest Euclidian distance from the client's scan (illustrated in Figure 4.5). Because there is some fluctuating interference from objects and radio sources, accuracy can be improved by using more sophisticated algorithms. A number of systems including RADAR, smooth out variations by averaging the locations of the K closest mapping scans in Euclidian space where a small $K$ of 2–4 has been shown to yield good results [24, 107]. An illustration of the $K$ nearest-neighbor algorithm is shown in Figure 4.6 for $K = 3$.

Other systems achieve additional improvements in accuracy by using Bayesian classifiers, radial basis functions, and other probabilistic techniques to estimate a client's location [21, 89, 155, 177]. These systems model the variation in signal strength as one or more probability distributions. (The histogram from Figure 4.2 could clearly be modeled fairly accurately with a single Gaussian.)

---

[1]We have described fingerprinting as "clients observing APs," as this is how most fingerprinting systems work. The original RADAR system actually had APs observing clients with symmetric algorithms.

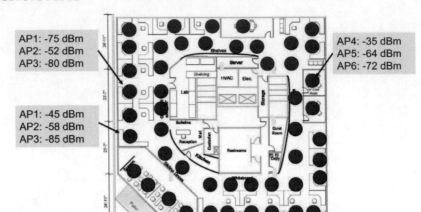

**FIGURE 4.3:** This map of an indoor office environment shows a reasonable set of locations in which to collect scans for a radio map. The three sample scans show that AP signal strength will vary across location and may potentially contain nonoverlapping sets of APs.

Given a radio map composed of signal strength distributions, these techniques can compute the likelihood that the client scan was performed at either a radio map location or an interpolated location in between mapping points. These likelihoods can then be combined to estimate the most probable client location. These probabilistic techniques have the advantage that, in addition to being able to estimate location more accurately, they can also approximate the error of their location estimate.

The accuracy of any location system based on 802.11 radio fingerprinting is dependent on a number of factors. Not surprisingly, system accuracy is strongly dependent on the density of the radio fingerprint map. The RADAR system was evaluated in a 1,000-m² office environment with 70 fingerprint locations, averaging 4 m distance between neighboring fingerprints and yielded a median localization error of 3 m (versus a median error of 16 m for random). Horus [176] and the commercial system Ekahau [40] report accuracies of 1 m or better using radio maps with 1–2 m between readings on average. On the other extreme, a 2006 study using cars with GPS to collect 802.11 data showed that with a mean distance of 10 m between fingerprints, it was possible to

| AP | Scan 1 | Scan 2 | Difference |
|-----|--------|--------|------------|
| AP1 | -70 db | -75 db | 5 db |
| AP2 | -50 db | -52 db | 2 db |
| AP3 | -90 db | -80 db | 10 db |

**FIGURE 4.4:** Two sample radio scans.

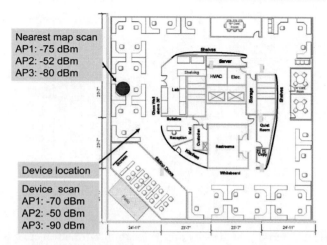

**FIGURE 4.5:** The nearest-neighbor algorithm estimates the device location to be the location of the mapping scan with the smallest distance in signal space.

estimate location with 15-m accuracy [24]. To factor out the effects of AP densities and physical differences in the deployment environment, King et al. varied the grid density of their radio maps and measured the effect on location estimation error [83]. Their results, shown in Figure 4.7, demonstrate the degradation in accuracy that can be expected as map density decreases.

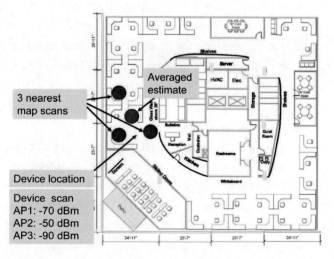

**FIGURE 4.6:** The $K$ nearest-neighbor algorithm estimates the device location to be the average of the $K$ (in this case $K = 3$) nearest radio map scans to the client's scan.

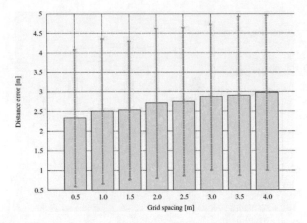

**FIGURE 4.7:** The effect of radio-map density on the location estimation error (both median and SDs of the errors are shown) [83].

Because location systems are often used by people carrying mobile devices, another factor to consider is the effect of human bodies on the accuracy of a fingerprinting location system. To mitigate the effect of bodies attenuating the 802.11 signals, RADAR's fingerprint maps were constructed with four fingerprints from each location, with the person doing the mapping facing each of N, S, E, and W. The rationale for taking these directional readings is to capture in the radio map both different locations and the different ways a person may occlude the signals. With all four readings included in the radio map, RADAR achieved median accuracy of 3 m, independent of the user orientation. On the other hand, when the fingerprint map was constrained to only the north-collected readings and testing was done while facing south, errors increased by more than 60%, showing that the effects of human occlusion on accuracy are indeed significant. Follow-on systems have varied in the number of mapping scans performed at each location varying from 1 in the case of maps generated from moving clients [24, 40] to as many as 100 [63, 176]. For more information about the effects of site survey density, readings per location, and AP density on the accuracy of fingerprinting systems, see the study by King et al. [82].

The strength of 802.11 fingerprinting is that it is a sampling technique: It does not need to understand where the APs are, what the physical environment looks like, or how 2.4-GHz signals propagate. It works by simply creating an approximate snapshot of the radio environment and later sampling that snapshot. This turns out to also be the weakness of this approach. First, the radio maps produced by the site surveys are brittle: If an access point is moved, the map needs to be recollected for the space within range of this AP, even if the move is as insignificant as from the top of a filing cabinet to the desk beside it. As a result, fingerprinting systems are best suited for spaces in which some degree of control over the radio infrastructure can be maintained such as office spaces, hospitals, and other similar buildings or small campuses. The second significant drawback is that fingerprinting does not scale well. Keeping a complete, up-to-date radio map for an office building or even a technology park might be fine, but is not feasible at the scale of countries, states, or even large cities. Studies have shown that fingerprint maps can be collected sparsely with a predictable loss in accuracy in exchange for density reduction [24]. We will see in the next section, however, that a better approach for doing location estimation with 802.11 over a wide area is to drop a sampling approach and rather to model radio signal propagation.

## 4.2    SIGNAL STRENGTH MODELING

The first wide-area deployment of an 802.11 location system was done as part of the Active Campus project on the University of San Diego campus in 2002 [58]. Because one of the goals of Active Campus was to offer location-based social networking services to students, a campus-wide indoor/outdoor location system was a requirement. While the campus did have pervasive 802.11 coverage, the campus was far too large to perform a fingerprinting site survey. One resource at the team's disposal, however, was the recent inventory the campus IT group had done recording the physical locations of the university's 1,000 access points. With this data set, the team developed algorithms to use this AP map to approximate a device's location given an 802.11 scan. Their system succeeded in achieving high coverage with accuracy in the wide-area of approximately 20 m [13]. The ability of Active Campus to use a relatively small amount of mapping data to provide high-coverage location estimation inspired a number of other model-based research [44, 95] and commercial [154] systems. In this section, we describe various systems, the approaches used to model signal propagation, and the ways to build the database of AP locations.

At their core, all these systems take advantage of the same principle: because radio signals propagate in predictable ways, a device's location can be predicted by modeling the propagation of the radio signals the device has observed [152]. While the models vary greatly, they all work by asking the same question: "Based on what I know about the access points just observed, where is the device most likely to be?" For example, Figure 4.8 shows the geographical coordinates for the AP captured in the scan results shown in Figure 4.1.

As a simplistic radio model, we might assume only that "802.11 APs have a range of 100 m." This model, combined with these AP locations, leads to the (perhaps incorrect) inference that the

| MAC | Latitude | Longitude |
|---|---|---|
| 00:0f:34:ab:0c:e0 | 43°39'39.95"N | 79°23'44.36"W |
| 00:0f:f7:0c:e9:c0 | 43°39'44.32"N | 79°23'47.33"W |
| 00:0f:f7:0c:4f:03 | 43°39'43.50"N | 79°23'37.86"W |

**FIGURE 4.8:** Geographical coordinates of sample APs.

client device is somewhere in the intersection of the three 100-m circles centered at the coordinates of the three APs as illustrated in Figure 4.9.

While simple, this example illustrates a few of the trade-offs in using a modeling-based approach. First, like all models, it makes simplifying assumptions that make the computation easier at the cost of decreased accuracy. In this case, we are assuming that all APs can be heard from the same distance in all directions, something that is almost never true. Second, we can see that even if we wanted to use a more sophisticated model, we could not do so if the AP database did not give us additional key parameters such as transmit strength or degree of obstruction that the model needed. This creates a strong coupling between the content of the AP database, the complexity of the radio model, and the resulting accuracy of the location estimation.

The Active Campus location system used an AP database only slightly more sophisticated than our example, storing the latitude, longitude, and floor number for each known access point. Active Campus used the observed signal strength from each access point to estimate the client's distance from that AP. The client's latitude and longitude were estimated by combining these distance estimates using an average that heavily weighted the distance estimate from the strongest AP. In their tests on campus, this technique had an average estimation accuracy of 22 m outdoors and 11 m indoors [13]. Bhasker et al. cite the increase in AP densities observed indoors as the reason for higher accuracy indoors, although it was also likely due in part to indoor obstructions effectively reducing AP ranges and shrinking the set of likely client locations [13]. The user's floor was always estimated to be the floor of the AP with the strongest received signal. Due to the heavy attenuation of concrete floors and ceilings, this technique yielded the correct floor 95% of the time in their tests.

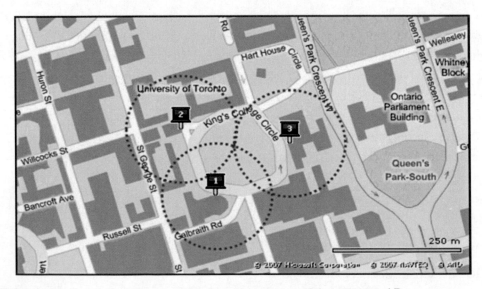

**FIGURE 4.9:** Predicted client location based on observation of three 802.11 APs.

One of the findings from the Active Campus deployment was that the participants wanted to continue using the system when they left campus at the end of the day. This created a problem, as the system's coverage did not extend beyond the range of the campus-managed network. The solution came in the form of the Place Lab system [95]. Like Active Campus, Place Lab employed a simple radio model based on an average of the latitude and longitude of the observed access points weighted by signal strength. The key contribution of Place Lab was the observation that an access point database could be quickly built using a technique known as "war driving" [95]. War driving is the practice of finding 802.11 networks by driving around in a car equipped with a GPS and an 802.11 device that is continually scanning and logging.[2] By post-processing the log, the GPS and 802.11 readings can be correlated to produce a cloud of locations in which a given AP was observed. By again using a model of signal propagation, this cloud of geo-tagged radio observations can be used to estimate the 802.11 access point's physical location. The observation of Place Lab was that war driving tools such as NetStumbler [118] and Kismet [84] that were primarily being used to allow people to find open networks could also be used to quickly build AP databases for 802.11 location systems. The strength of this approach is that war driving will map all observable access points, regardless of whether they are in a school, a business, or a private residence and independent of whether their traffic is encrypted or not. This allowed Place Lab to provide coverage beyond that of a managed network, making it the first system to potentially offer city or country scale 802.11-based location. Experiments with Place Lab yielded accuracy numbers very similar to those of Active Campus. In the neighborhood with the sparsest AP distributions, Place Lab has a median accuracy of 23 m, while in the densest, it had accuracies of 15–20 [95].

## 4.2.1    Constructing the AP Database

Active Campus and Place Lab illustrate two popular means of constructing an access-point database. The advantage of the Active Campus approach is accuracy: Knowing an AP's location on a floor plan or a building schematic allows someone to estimate the AP's absolute location within a meter of its true location. The downside of this approach, as we've already mentioned, is that it really only works when a network manager knows the location of their networking hardware. (Note that this does not completely eliminate the possibility of achieving wide area or high coverage with this technique. Cities deploying metropolitan-scale WiFi programs or even national "hotspot" providers like T-Mobile likely know the location of their APs.) War driving, on the other hand, allows vast

---

[2]The term "war driving" alludes to the term "war dialing" the act of setting a modem to sequentially scanning phone numbers to find other modems. The term "war dialing" came from the 1983 movie War Games which popularized this technique for finding modem-accessible networks.

swaths of terrain to be quickly mapped at the cost of reduced accuracy in the AP location estimates. The errors in AP placement come from a variety of sources: first, GPS is being used to record the location of radio observations, and GPS itself has an 8- to 10-m median error. This is compounded by the error introduced by the modeling used to predict AP location from the set of observations. A comparison of war-driven AP locations to the true location of five APs in a residential area found a median error of 26 m [96]. A larger study of over 500 APs on the Dartmouth campus found a median error of 40 m for war-driven AP estimates [82]. A third source of error comes from that fact that all of the popular war-driving programs ignore altitude and estimate AP location in just latitude and longitude. While this may not introduce significant error for APs in one- or two-story buildings, Place Lab found that it had a significant effect in downtown areas with tall buildings [95]. Finally, due to the unmanaged, organic nature of the collection of APs being mapped, it is almost certain that APs are being deployed, decommissioned, or moved on a regular basis in the mapped area. This gives the data from a war drive an unpredictable, temporal fragility that a map of centrally controlled infrastructure could determine and even control. An illustration of this was provided by Cheng et al. [24] during their characterization of 802.11 location performance in the Seattle metropolitan area. In a follow-up war drive of an area that they had previously mapped, 4 months earlier, they found a 50% change in the set of APs in the area. This example is likely on the extreme end of what could be expected for turnover, as the two war drives spanned the start of a school year in a student-heavy college neighborhood. Regardless, it serves to illustrate the importance of Periodically refreshing the AP map data.

Despite these drawbacks, war driving has been far and away the largest source of AP data for location estimation. The commercial location system by Skyhook Wireless and Navizon employ nationwide war driving in the United States to build their AP maps [116, 154]. In the public domain, the user-contributor war-driving repository "Wigle" contains the location of over 11 million APs at the time of writing [174]. For example, Figure 4.10 shows maps of Wigle.net's AP coverage for the continental United States and for Manhattan Island.

Chandrasekaran et al. have investigated the feasibility of populating an AP database given the postal address of the access point [22]. The study does not address how the postal address of APs would be collected, but rather showed that were such data available; AP locations could be estimated reasonably using geocoding services like Yahoo! maps. In a residential neighborhood in New Jersey, their study showed that, given the correct postal address, the AP location was estimated with an average error of around 100 m.

Finally, LaMarca et al. have shown that it is possible to estimate AP locations solely using unlabeled 802.11 traces given the known location of a small percentage of the APs [96]. A constraint-based graph algorithm uses temporal correlations between observations of known and unknown APs to estimate the location of the unknown APs. Their experiments showed that using self-mapping with a seed set containing as little as 10% of the APs is possible to build an AP data-

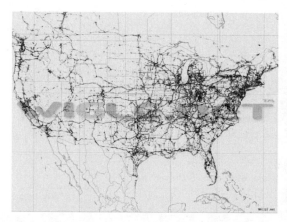

**FIGURE 4.10:** Maps depicting the locations of some of the 11 million APs that are stored in the user–contributor wigle.net database. Both the map of the continental United States (left) and Manhattan Island (right) indicate a correlation between AP density and population. Source: http://wigle.net.

base that provides almost 90% coverage. Their experiment also showed that for an up-to-date radio map, self-mapping can add new beacons with nearly the same accuracy as war driving.

## 4.3   PRIVACY CONSIDERATIONS

802.11 location systems have been architected in various ways and this has an impact on the privacy guarantees they offer their users. The privacy of 802.11 location systems can be excellent if, like GPS, it is implemented entirely with passive radio reception and client-side computation. This client-side approach is used by Place Lab, Ekahau, and others. Other systems have, for various reasons, implemented some of the functionality in the infrastructure, lessening a client's privacy guarantees. The original RADAR system was actually implemented by having APs listen for packets from clients and then pool their observations to make an estimate of the client's location. Skyhook Wireless and Active Campus run a small client-side program that performs the network scan, the results of which are shipped to a central server that performs the location estimation. On the upside, this allows the client software to be minimal and the AP database to be both centrally maintained and safe guarded. On the downside, it both requires the client to have network connectivity to do location estimation and compromises the user's privacy.

## 4.4   IMPROVEMENTS AND VARIANTS

In places with 802.11 network coverage, a signal fingerprinting approach should provide near 100% coverage with 1- to 3-m accuracy [63, 107, 176]. To give the reader a sense of the distribution of location errors typical of a fingerprinting system, Figure 4.11 shows an error histogram and a CDF for the Locadio system by Krumm and Horvitz [89]. In these same locations, a signal-modeling

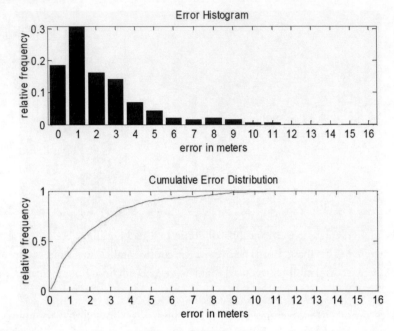

**FIGURE 4.11:** The error distribution and error CDF for the Locadio 802.11 location system [89].

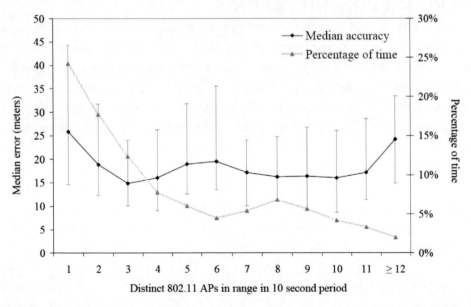

**FIGURE 4.12:** The median error of the Place Lab system as a function of the number of distinct observable APs. The figure also shows how often a given number of distinct APs are detected. Mapping and testing were performed while driving in the Seattle Metro area in the US. Error bars show the first and third error quartiles [95].

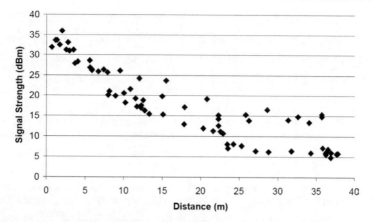

**FIGURE 4.13:** This graph demonstrates that indoors the strength with which clients observe APs drops off with distance from the AP [8]. © 2000 IEEE.

approach, using a manually built or war-driving AP map, can expect to predict location with 10- to 25-m median accuracy [13, 24, 96]. Again, to give the reader a sense of expected performance, Figure 4.12 shows the distribution of errors for Place Lab as a function of the number of known APs [95].

Various approaches have been used to improve the accuracy of 802.11 location systems. In their original RADAR paper, Bahl et al. demonstrated that, even indoors, there is a strong correla-

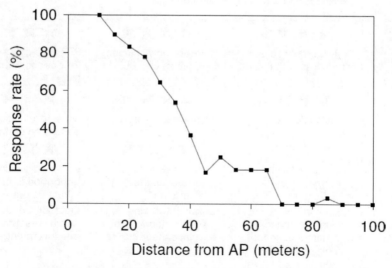

**FIGURE 4.14:** Response rate for an AP as a function of distance from that AP. Response rate is defined to be the percentage of radio scans in which an AP was observed [24]. © 2005, ACM, Inc. Reprinted by permission.

tion between distance from an AP and observed signal strength (shown in Figure 4.13). All of the systems we have mentioned take advantage of this and use the received signal strengths from APs to refine their location estimates. Cheng et al. showed, however, that the percentage of beacon frames received by a client from a given AP (which they call "response rate") could also serve as a good

| Technology | 802.11 signal-strength fingerprinting (e.g., RADAR) | 802.11 signal-strength modeling (e.g., Place Lab) | 802.11 proximity (e.g., GUIDE) |
|---|---|---|---|
| Accuracy | ★★★☆ 2D coordinates with 1-3 m median accuracy | ★★☆☆ 2D coordinates with 10-20m median accuracy | ★☆☆☆ Location accuracy dependant on AP density |
| Coverage | ★★☆☆ Building to campus scale. Requires 802.11 coverage and radio map. Best accuracy achieved when 3+ APs are visible. | ★★★☆ Areas with 802.11 coverage and radio map. Best accuracy achieved when 3+ APs are visible. | ★★★☆ Anywhere with 802.11 coverage and an AP, location map. |
| Infrastructure cost | ★★☆☆ No additional infrastructure is needed beyond 802.11 APs. Creating radio map is time intensive and new/moved APs require remap. | ★★★☆ No additional infrastructure is needed beyond 802.11 APs. Creating radio maps is less work than for fingerprinting. | ★★★★ No additional infrastructure is needed beyond 802.11 APs. |
| Per-client cost | ★★★★ Software-only solution for devices with 802.11 NICs. | ★★★★ Software-only solution for devices with 802.11 NICs. | ★★★★ Software-only solution for devices with 802.11 NICs. |
| Privacy | ★★★★ when localization is performed on the client. ★☆☆☆ when localization is performed in the infrastructure. | ★★★★ when localization is performed on the client. ★☆☆☆ when localization is performed in the infrastructure. | ★★★★ when localization is performed on the client. ★☆☆☆ when localization is performed in the infrastructure. |
| Well-matched use cases | Asset and personnel tracking in indoor environments, indoor mapping/navigation/tour guides | Social networking, tour guides, indoor/outdoor navigation/tour guides, fitness/activity tracking | Outdoor tour guides, nearby resource advertisement, activity tracking |

**FIGURE 4.15:** A summary of the performance characteristics of 802.11 location estimation techniques.

metric for refining location prediction [24]. Figure 4.14 shows how response rate fell with distance from an AP in their experiments. Many systems have also used temporal smoothing techniques such as particle filters [36] to track users across many radio scans and, thus, reduce errors (Particle filters and sensor fusion are discussed in detail in Chapter 7).

While nearly all 802.11 location systems use either signal strength fingerprinting or modeling to estimate a device's coordinates, other approaches have been tried. Simple proximity systems use the 802.11 access point, a device is associated with, as an indication of its rough location. This approach was used successfully in the GUIDE system, a location-aware walking tour of Lancaster, UK [26]. Krumm et al. have developed a sophisticated system called NearMe based on coordinate-free proximity [88]. By collecting and comparing streams of fingerprints, NearMe is able to construct a topology of the places people go. While no coordinate-based information is collected or predicted, NearMe is able to say which people and things are near each other and is even able to estimate transit time between nearby places based on past data.

While time-of-flight ranging has been proposed as a way to locate 802.11 clients, this approach has two factors working against it. The first, and probably most significant issue, is that RF ranging requires specialized hardware in both the clients and the access points to enable the transmission times to be accurately measured. This undermines the biggest advantage of 802.11 location systems—that no extra hardware or infrastructure is required. Beyond that, 802.11 ranging is hard to make accurate, as the relatively narrow frequency bands used by 802.11 make multipath hard to detect and eliminate. Despite the short distances involved, ranging 2.4 GHz error in indoors spaces is around 3 m [79], offering little advantage over good signal fingerprinting techniques.

Finally, two systems have attempted to combine the ease of deploying a signal modeling technique with the accuracy of a fingerprinting technique. To improve their ability to predict room names from their location estimates, Active Campus includes a correction process whereby a user can, using a graphical map, inform the system exactly where they are at that time. Active Campus then collects a RADAR-style fingerprint and stores it in a fingerprint cache for that user. These manual corrections allow Active Campus to employ a hybrid location scheme, wherein room prediction accuracy rose to 90% when there is a closely matching fingerprint in the user's cache. Because people tend to have a relatively small number of important places in which they spend much time, the number of corrections per user should be small. A system by Letchner et al. achieves similar functionality by using a signal propagation model which can be trained using a mix of fingerprint scans and AP characteristics [99]. This unified system allows for a smooth transition between low accuracy predictions based on a few AP characteristics to more accurate prediction where detailed site surveys have been performed. We close with a table shown in Figure 4.15 summarizing the performance characteristics of the fingerprinting, modeling and proximity approaches to 802.11 location estimation.

· · · ·

CHAPTER 5

# Cellular-Based Systems

The development of location systems based on mobile phone technology was originally driven by the U.S. Federal Communication Commission E911 mandate and its European Community counterpart E112 to locate mobile phone calls to assist in the delivery of emergency services. In addition, the wide adoption and ubiquitous connectivity of cellular phones makes them tempting platforms for the delivery of location-based services, such as advertising, recommendation systems, and gaming. A localization system based on cellular signals, such as GSM or CDMA, has the key advantages that it leverages the phone's existing hardware and can potentially provide location estimates anywhere voice service is available.

A number of mobile phone-based location systems have been developed, and they can be grouped into four categories: cell ID-based approaches, methods based on radio propagation modeling, assisted GPS, and surveying techniques based on radio fingerprinting. As could be expected, each type of location system strikes a different trade-off between ease of deployment, coverage, and accuracy. The rest of this chapter discusses mobile-phone-based location systems following the above taxonomy. We conclude the chapter with a comparison between the different types of systems and a brief discussion of standardization efforts.

## 5.1    CELL ID-BASED

A mobile-phone base station is typically equipped with a number of directional antennas that define sectors of coverage or *cells*, each of which is assigned a unique cell ID. Cell ID-based location is a simple technique where the position of the mobile phone is estimated based on the ID of the cell currently providing service to the device.

Cell ID-based location is usually implemented on the network side, and its key advantage is that it works for all phones, as no handset modifications are required. Accuracy depends on the size of the cell, ranging from 150 m for microcells in urban cores to 30 km for cells in rural settings. This level of accuracy may be sufficient for some applications, such as weather and traffic reports, but falls short of the requirements for other application such as street navigation and the E911 guidelines, which require that a cell phone handset be localized within 50 m 66% of the time. Accuracy can be improved by including in the position calculation the round-trip time (RTT) [15].

Laasonen et al. [92] is an example of a cell-ID based system implemented purely on the handset without network cooperation. They monitor the ID of the GSM cell the handset receives service from and uses the transition between cells to recognize the common places that a user goes to. Because Laasonen's system does not attempt to estimate absolute location, but rather assigns symbolic names like Home and Grocery Store to locations, the system does not need to know the coordinates of the cell towers. The privacy improvement in this "handset-only" approach is marginal, as current protocols require the mobile phone to register with the cell tower to obtain the cell ID.

## 5.2    RADIO MODELING APPROACHES

A variety of location systems based on modeling of GSM and CDMA radio signals have been developed. Unlike the 802.11 systems we saw in the last chapter, most of the radio modeling systems for GSM and CDMA are based on time-of-flight measurements rather than signal strength. It is worth noting that while the underlying techniques used for computing ranges to cell towers depend on the specifics of the GSM or CDMA physical layer, the positioning algorithms used by GSM and CDMA are in fact very similar.

The main radio modeling positioning methods provided by GSM and CDMA systems are based on Time Difference of Arrival (TDOA) trilateration [19, 38]. In TDOA, handset position is estimated by intersecting hyperbolic lines derived by taking differences between time measurements from pairs of base stations. Figure 5.1c illustrates the intersection of two hyperbolic curves derived from measurements to three base stations.

TDOA requires base stations to be tightly synchronized. While this is not an issue for CDMA networks, which follow a synchronous protocol, meeting this requirement in GSM networks requires the deployment of additional network elements to determine the clock difference between base stations. This information is then provided (by the base station) to the handset who uses it to calibrate the measurements from neighboring base stations.

TDOA can be implemented either on the network or the client. In the network-based approach, three or more base stations compare observations to determine the time differences at which handset transmissions are heard at the cell towers. In the client-based implementation, the handset measures the time differences in the arrival of training sequences or pilot symbols transmitted by three or more base stations. The network-based approach has the advantage that it does not require modifications to the handset hardware, and as a result, it works for existing, deployed handsets. On the other hand, the client-based approach provides a higher degree of privacy to users, as time-difference is measured and the location estimated on the user's local device. Overall, TDOA accuracy ranges between 50 and 500 m depending on interference, system geometry, and multipath effects. Handsets in close proximity to a base station may suffer from interference issues where the strong signal of the nearby base station prevents the handset from hearing the transmission of neighboring nodes. This problem is more pronounced for CDMA systems, where different towers transmit on the same fre-

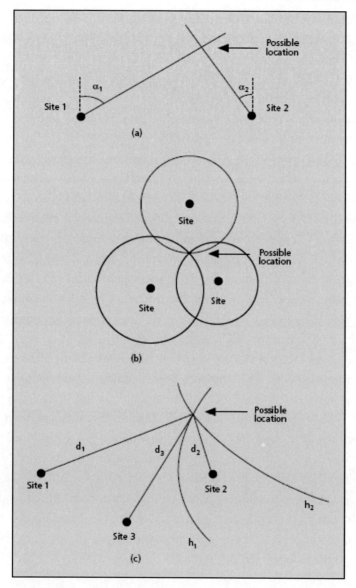

**FIGURE 5.1:** Three location algorithms based on radio signal modeling: (a) angle of arrival; (b) time of arrival; (c) time difference of arrival [178]. © 2002 IEEE.

quency, than on GSM networks, where towers are allocated different frequencies. To alleviate these situations, base stations stop transmission for short idle periods that enable the handset to measure other neighboring nodes. As is the case for GPS, location accuracy is a function of system geometry. For example, a set of base stations positioned on a straight line along a rural highway would produce a low quality position fix for devices moving along the road. Finally, multipath effects, resulting from signal reflecting on obstacles, degrade the accuracy of the technique by lengthening the perceived distance to the base station.

In addition to TDOA, GSM, and CDMA also provide positioning methods based on angle or arrival (AOA) and time of arrival (TOA). AOA is a method that determines position by electronically steering a directional antenna or an antenna array. Handset position is determined using triangulation, by intersecting a minimum of two directional lines of bearing as shown in Figure 5.1a. The most significant drawback of AOA is that it requires specialized antennas and receivers, which limits this approach to network-based implementations. On the other hand, AOA has the advantage that it does not require system-wide synchronization and is more robust to multipath propagation effects. In TOA, the handset position is determined by intersecting range circles as shown in Figure 5.1b. TOA systems can be implemented at either the network or the client. A significant downside of TOA systems is that the measuring device has to have accurate knowledge of the time of transmission to determine the signal propagation time [111]. We note that TDOA systems get around this requirement by taking the difference between measurements. Gustafsson and Gunnarsson provide an analysis of the fundamental performance limitations of different geometrical location methods based on cellular signals [61].

GSM and CDMA location systems based on the measurement of received signal strength have also been developed [159]. This approach is less common than time-based approaches, as it typically achieves lower performance [173]. One example is the Place Lab system [95], which complements their 802.11 signal models with a GSM model that predicts tower distance based on received signal strength. In Place Lab, GSM cell tower locations are estimated using war-driving tools similar to those used to estimate 802.11 base station locations. In an evaluation in three Seattle neighborhoods, Place Lab achieved median accuracy between 100 and 200 m depending on building and tower density; more densely populated neighborhoods, which tend to have a higher concentration of base stations, had better accuracy. The handset-based approach enables location determination without the need for cooperation from the network operator but requires the mobile to store a database of tower locations.

## 5.3 ASSISTED GPS

Assisted GPS or *A-GPS* [34] is a hybrid positioning technique that combines the Global Positioning System (GPS) and the mobile phone network. GPS is a satellite-based location system that achieves accuracies of 10 m or better when the mobile receiver has a clear view of the sky and can receive a signal from four or more GPS satellites. (For more information on GPS, see Chapter 2.) Integrating GPS receivers on mobile phones presents serious challenges: GPS location estimation uses significant computational resources and power, and receivers in dense urban environments often do not have an unblocked view of the sky and are unable to detect the four GPS satellites required to estimate location. In addition, an uninitialized GPS receiver with limited or no knowledge of the state of the GPS constellation can take several minutes to acquire a GPS fix, raising usability and safety issues [178].

A-GPS is a differential GPS technique that enables position determination when fewer than four GPS satellites are visible on the mobile handset. A-GPS supplements the observations of the GPS receiver on the mobile handset with measurements collected by a second nearby reference GPS receiver positioned in a known location with a clear view of the sky. While not as accurate as a true GPS location fix, the supplementary signals allow the handset to estimate its location, even when as few as one satellite can be heard. A-GPS has the additional advantage that it reduces the time from when the handset is turned on to the initial location acquisition by relaying, over the cellular link, the current state of the GPS constellation. Finally, like other differential GPS techniques, A-GPS can improve location accuracy by leveraging the measurements collected by the reference receiver to correct for errors in the broadcasted satellite coordinates and biases introduced by atmospheric effects.

Figure 5.2 illustrates the A-GPS architecture [34]. The reference GPS receiver can be co-located with the base station or at a dedicated A-GPS server that serves a group of base stations.

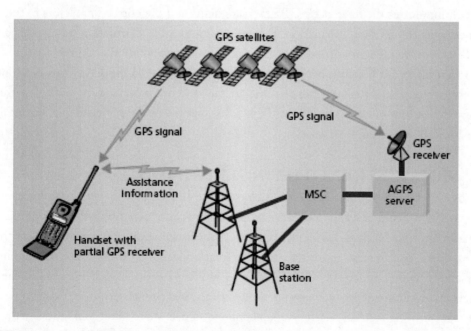

**FIGURE 5.2:** Assisted GPS. The position of the mobile handset is determined by combining measurements collected by a GPS receiver on the handset with those collected by a reference GPS receiver at a known location. MSC in the figure stands for mobile switch center [34]. © 2001 IEEE.

**TABLE 5.1:** QUALCOMM's gpsOne performance for different combinations of GPS satellites and range measurements to cellular base stations

|  | ERROR (m) | |
|---|---|---|
|  | 67% | 95% |
| 4 satellites | 11 | 18 |
| 3 satellites + 1 base stations | 19 | 35 |
| 2 satellites + 2 base stations | 88 | 253 |
| 1 satellites + 3 base stations | 139 | 345 |

In a network-based A-GPS implementation, the handset relays its GPS observations to the base station, which then determines the handset's location. This reduces handset resource requirements by offloading much of the computation to the infrastructure, but creates more of a privacy issue as the handset's movements can be tracked in great detail. In a handset-based approach, the base station communicates the reference's GPS measurements and the correction factors to the handset which performs the location calculation. By performing location calculation locally, this approach preserves user privacy, but does so at the cost of additional computation and power usage.

Advanced A-GPS solutions, such as QUALCOMM's gpsOne [158] can combine range measurements to cellular base stations and satellite measurements to improve location performance when fewer than four satellites are visible. Table 5.1 shows the performance of gpsOne for different satellite and cell tower combinations.

## 5.4   SIGNAL STRENGTH FINGERPRINTING

Signal strength fingerprinting is an empirical location technique that involves a "training" or "mapping" phase in which a radio map of the environment is constructed by collecting a series of fingerprints in multiple locations. A radio fingerprint reflects some property, commonly as the signal strengths, of a group of radio sources that are heard at a specific location. Once the training phase is complete, a client can determine its location by looking for the closest matches of the current measurement to the set of measurements collected in the training phase.

Fingerprinting systems based on mobile phone signals share many similarities with the systems based on 802.11 we described in Chapter 4. There are, however, differences due to the underlying technologies on which these systems are built. Due to higher coverage, cellular fingerprinting has the potential to work in more places than 802.11 fingerprinting. Moreover, because cellular towers are dispersed across the covered area, a cellular-based localization system would still work in situations where a building's electrical infrastructure has failed. The significant expense and complexity of cellular base stations result in a network that evolves slowly and is only reconfigured infrequently. While this lack of flexibility (and high configuration cost) is certainly a drawback for the cellular system operator, it creates a stable environment, and radio maps for cellular systems will degrade at a slower rate than those based on 802.11. Finally, due to shorter range, 802.11 fingerprinting will be more accurate than cellular fingerprinting given the same number of radio sources.

In principle, both GSM and CDMA systems are equally amenable to signal fingerprinting. In practice, the radio signal information, such as signal strength or time delay needed to implement fingerprinting, is more readily available to user-level programs on GSM equipment than on CDMA. As a result, research on mobile phone-based location fingerprinting has focused almost exclusively on GSM.

Signal strength information is not uniformly available across GSM receivers. Many GSM phones make no network information available to user-level programs. Others provide the ID and observed signal strength of the current cell providing the handset with coverage. The phone-hacking community has shown that by directly observing the phone's memory, it is possible to learn ID and observed strength of six nearby cells. The Sony Ericsson GM28 GSM modem (shown in Figure 5.3) is an interesting device, as it is inexpensive and offers considerable information about the current GSM environment. It actually provides two different interfaces for accessing information: one based on cells and the other on channels. The cell interface reports the ID, signal strength, and associated channel for up to seven nearby cells. The channel interface provides the signal strength for up to 35 GSM channels! In practice, the six strongest signals in the channel interface will be six of the cells from the cell interface. While providing information about more cells, the channel interface reports only signal strength and not cell IDs. The lack of a standardized interface for accessing signal strength information constitutes an important obstacle to the research and eventual deployment of fingerprinting systems.

While GSM employs transmission power control both at the base station and the mobile device, the base station's main control channel (known as BCCH) is always transmitted at a full and constant power to allow mobile stations to compare the signal strength from neighboring cells. Observations of these BCCH transmissions make up the measurements used in GSM fingerprinting systems.

The performance of GSM fingerprinting for localization in outdoor environments was first evaluated by Laitinen et al. [94]. Their system used sparse fingerprints from the six strongest cells to achieve 67th percentile accuracy of 44 m. Varshavsky et al. [166] observed that it is possible to significantly increase the accuracy of GSM fingerprinting in indoor environments by using wide signal-strength fingerprints. The wide fingerprint includes the six strongest GSM cells and readings of up to 29 additional GSM channels, most of which are strong enough to be detected, but too weak to be used

Sony Ericsson GM 28

**FIGURE 5.3:** Sony Ericsson GM 28 GSM modem.

for efficient communication. They showed that the higher dimensionality introduced by the additional channels dramatically increases localization accuracy. Figure 5.4 plots the median localization error as a function of the number of channels used in the fingerprint. Overall, their system achieves median within floor accuracy of 4 m in large buildings [122] and is able to identify the floor correctly in up to 60% of the cases and is within two floors in up to 98% of the cases in tall multi-floor buildings [167].

Varshavsky et al. obtained their wide fingerprints using the Sony Ericsson GSM model shown in Figure 5.3. Chen et al. [23] showed that similarly wide GSM fingerprints could be obtained if

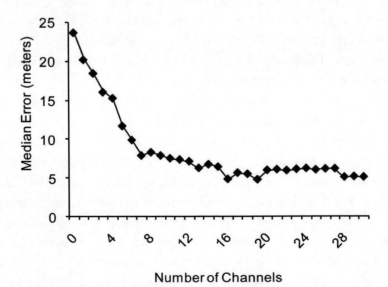

**FIGURE 5.4:** Location error as a function of fingerprint size [122].

standard handsets were able to simultaneously observe towers from the two or more GSM network operators that often compete in an area.

As is the case for 802.11 systems, the key drawback of GSM fingerprinting systems is the need to collect a large number of measurements to train the system. One way to ameliorate this problem is to use a radio propagation model to derive a detailed radio map from a much smaller set of measurements. Chen et al. [23] developed a system that models signal propagation using Gaussian Processes, which are nonparametric models that estimate Gaussian distributions over functions based on the training data. Because the Gaussian processes algorithm models signal strengths with continuous functions, this technique works better in more open environments, such as residential areas. In high-density downtown environments, however, the technique fails to capture the sharp changes in signal strengths due to obstructions. Similarly, Zhu and Durgin [179] developed a system that uses the Hata signal propagation model to generate the radio map.

## 5.5 STANDARDIZATION EFFORTS AND DISCUSSION

Standards for the deployment of location technologies by cellular providers have been published by the Third Generation Partnership Project (3GPP) and the 3GPP2, which concentrate on the GSM family of protocols (including W-CDMA) and CDMA, respectively [178]. These specifications call for the implementation of three location services: Cell-ID, TDOA, and A-GPS. Each of these types of location services strikes a different trade-off between ease of deployment, coverage, and accuracy. Cell ID provides the lowest accuracy, but because it does not require additional hardware on the network side, it is the one method that is universally supported by all network providers. TDOA provides higher accuracy, but because it requires changes to the network, the handset, or both, it is not yet universally supported. A-GPS provides accuracy in the tens of meters, but is still only available on a small fraction of mobile handsets [57].

A-GPS and TDOA complement each other to a large extent. A-GPS is expected to perform best in rural and suburban environments where few obstructions provide for good satellite visibility. In this environment, however, low base station density leads to poor TDOA performance. Conversely, in downtowns and other dense urban environments, TDOA has higher coverage than A-GPS, as it works (albeit at reduced accuracy) indoors and in other obstacle-rich environments.

Whereas location systems based on Cell ID, TDOA, and A-GPS are commercially available, mobile phone fingerprinting systems are still at the stage of experimental prototypes. Radio fingerprinting systems can achieve high accuracy, but the need to collect training measurements limit their coverage. While these systems do not require additional hardware, the lack of a standardized interface for accessing signal strength information represents an important obstacle to their commercial deployment. Figure 5.5 shows a table summarizing the performance characteristics of these four approaches to cell-based location estimation.

| Technology | GSM signal-strength fingerprinting | GSM TOF and signal-strength modeling | GSM/CDMA proximity | Assisted GPS (A-GPS) |
|---|---|---|---|---|
| Accuracy | ★★★☆<br>2D coordinates with 4 m median accuracy in dense cell environment | ★★☆☆<br>3D coordinates with 100 - 200 m accuracy | ★☆☆☆<br>Accuracy dependant on cell tower density (150 m – 30 km) | ★★☆☆<br>3D coordinates with 10 -150 m accuracy depending on number of GPS satellites visible |
| Coverage | ★★☆☆<br>Building to campus scale. Requires cell network coverage and radio map. Best accuracy when 3+ cells are visible | ★★★☆<br>Areas with GSM coverage and radio map. Best accuracy when 3+ cells are visible | ★★★★<br>Anywhere with cell coverage and cell-to-location map | ★★★☆<br>Outdoors with 4+ GPS satellites or indoors with cell network support + view of 1+ GPS satellite |
| Infrastructure cost | ★★★☆<br>No additional infrastructure is needed beyond cell network. Creating radio map is time intensive | ★★★★<br>No additional infrastructure is needed beyond cell network and map of tower locations | ★★★★<br>No additional infrastructure is needed beyond cell network and map of tower locations | ★★★☆<br>Beyond GPS constellation, requires deployment of fixed GPS receivers |
| Per-client cost | ★★★★<br>Software only solution | ★★★★<br>Software only solution | ★★★★<br>Software only solution | ★★★☆<br>GPS antenna and chipset required for handset |
| Privacy | ★★☆☆<br>Even if location is computed on client device, the network still tracks a handset's associated cell | ★★☆☆<br>Even if location is computed on client device, the network still tracks a handset's associated cell | ★★☆☆<br>Even if location is computed on client device, the network still tracks a handset's associated cell | ★★☆☆<br>Even if location is computed on client device, the network still tracks a handset's associated cell |
| Well-matched use cases | Asset and personnel tracking in indoor environments, indoor mapping/navigation/tour guides | Social networking, emergency response, neighorhood-scale information access, fitness tracking, outdoor mapping / navigation | Regional information access (weather, traffic, etc.) | Emergency response, indoor/outdoor information/tour guide services, personnel/pet tracking, activity tracking, gaming |

**FIGURE 5.5:** A summary of the characteristics of common GSM/CDMA location estimation techniques.

CHAPTER 6

# Other Approaches

Thus far, we have focused on describing the most widely deployed, general-purpose approaches for locating devices, objects, and mobile users. While the location technologies we have described thus far vary in cost, coverage, and accuracy, all are used in commercial location systems, and all have reached wide-scale adoption in one or more application niches. There are, however, many other location technologies that have not gotten beyond the research laboratory or the prototype stage. There are others, too, that have been largely perfected, but due to small coverage or exceptionally high cost have not moved beyond a few highly specialized niches. As many of these systems have novel features and employ interesting algorithms, we spend this chapter touching briefly on these other technologies. These include the most accurate system we will discuss, with 1 mm average error, and the least, with almost 100 km average error. A number of systems eliminate the need for people to carry beacons or listeners and instead use a person's visual appearance, noise profile, or ground reaction forces to track them through an environment. In addition to the GPS, 802.11, and cell-phone-based system we have described, a legion of other radio-based systems have been developed based on both standard and custom protocols ranging in spectrum from 33 kHz up to 7 GHz.

## 6.1    INSTRUMENTED SURFACES

One of the most significant drawbacks of the systems that we have presented thus far is that they are only able to track devices, objects, and people that are participating in the protocols required by the location system, and the system is blind to those that do not. While this offers a clear privacy advantage and is less of an issue for devices that are rarely if ever separated from their radios, it creates problems when everyday services like lighting and heating adjustment are reliant on people wearing smart badges [107]. An alternative is to base location estimation on something physical and fundamental, such as instrumenting the floor to track people by their weight. Schmidt et al. do this by segmenting a room's floor into large 4-m² panels (180 × 240 cm) that rest on load sensors, one under each corner [149]. They show that, by comparing the change in load across the four corners, they can estimate the location of the center of the force with 10-cm accuracy. Using periods of static load, the system is able to "auto-tare" and discount the weight and distribution of stationary objects like furniture. While accurate and free of instrumentation for users, such a system has two obvious drawbacks. First, it is only able to measure the location of a single person on a panel at a time, and second, the system is anonymous, and while it knows someone's location, it does not know who they are.

Both of these issues are addressed in the Smart Floor system [121], which employs a smaller floor segment (50 × 50 cm), which carries a lesser risk of being occupied by multiple people. The Smart Floor does not use the load sensor data to estimate inter-tile location, but rather uses the ground reaction forces it measures as a means to estimate the identity of the person stepping on it. Figure 6.1 shows a time series graph of the reaction force of a footfall and points out the features the system uses to identify users. Their experiments showed that from a pool of 15 users, the system is able to correctly pair a footfall with a person 93% of the time.

Another notable instrumented surface is the Magic Carpet system that employs a mesh of piezoelectric polymer placed underneath a carpet [126]. By measuring along which rows and columns electricity was induced, the system can estimate where a person is standing. The most unusual of the instrumented floors is the CarpetLAN system, which uses the conductivity of the human body to serve as a wired network between a worn or carried device and LAN-enabled floor tiles [50]. As these floor tiles have distinct IDs, a user can be located by which floor segment is routing their data.

Systems based on load sensors placed in the floor are likely to be robust, unobtrusive, and low-maintenance, and whereas some of these systems are limited in their handling of multiple occupancy situations, they can correctly detect occupancy and thus still work well for occupancy-based services. The main downside of this class of systems is that they cannot be easily retrofit into an

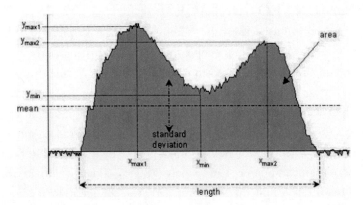

**FIGURE 6.1:** A graph of the ground reaction force of a floor tile in the Smart Floor. The features shown in this profile were able to identify the person stepping on the tile from a group of 15 people with 93% accuracy [121]. © 2000, ACM, Inc. Reprinted by permission.

existing home or office space, making this approach only suitable for new construction or structurally remodeled spaces.

## 6.2    VISION

Images from cameras have obvious potential for locating and tracking, as vision is the primary sense that humans use for this same purpose. It is also one of the few approaches that do not require instrumenting or tagging the entity to be located. We briefly review the state of the art in vision-based location, both using mobile and fixed cameras, and describe how tags have been used to assist in identification and tracking. Finally, we discuss practical considerations related to the deployment of vision-based location systems.

### 6.2.1    Mobile Cameras

The use of a mobile camera to track the location of the camera bearer has primarily been used for robotic navigation [32, 87]. These systems use a single mono or stereo camera, mounted forward, to identify landmarks or features such as doorways and thresholds in the environment [74, 143]. By tracking notable features across multiple frames, features can be triangulated to permit a 3-D location to be estimated. The current state of the art in illumination and viewpoint invariant feature extraction is the Scale Invariant Feature Transform or SIFT algorithm [104]. Figure 6.2 shows an example of SIFT feature recognition in an outdoor scene. One of the keys of SIFT is the speed with which features can be matched in target images.

Real-time egocentric tracking algorithms unfortunately do not translate well from robots to humans. Less information is available, as there is no odometry (wheel turn information), and the video quality is decreased, as people do not exhibit the smooth, slow movement patterns that robots do. Single-frame camera-based location algorithms have been developed for mobile cameras [49]. Hile and Borriello, for example, have used SIFT features extracted from camera phone images to do augmented reality overlays to provide directions (see Figure 6.3) or other location-sensitive information [72]. Their system uses 802.11 fingerprinting to reduce the set of SIFT features to be searched. Despite this optimization, it takes their system 10 s to estimate location and heading.

### 6.2.2    Fixed Cameras

A more popular approach for vision-based location estimation is to embed fixed cameras in the environment and to track the location of the people that appear on camera. Such systems have been used both indoors and outdoors for security and surveillance [27], as well as in "smart room" deployments that want to track people without requiring them to carry devices or special badges [17]. An example of outdoor people tracking is shown in Figure 6.4, and an example of indoor object tracking is shown in Figure 6.5. Because these systems are primarily interested in the parts of the

**FIGURE 6.2:** An example of SIFT feature recognition. Images of previously learned landmarks are shown upper left. The lower large image shows the scale and viewpoint invariant recognition of these landmarks in the upper large image [104]. © 1999 IEEE.

**FIGURE 6.3:** Vision-based location systems have been used with camera phones to overlay navigation information on the camera image [72].

**FIGURE 6.4:** Sample images of the W4 system using head, torso, and limb detection to track people with mono camera [65]. © 2000 IEEE.

images that are moving, they do not need background elements like furniture, windows, and doors. This allows them to do background subtraction and focus on the dynamic parts of the scene, which enables their algorithms to run much faster (typically real-time or faster [28]) than those from the last subsection.

In general, these systems track objects using a bottom-up approach, in which blobs of color, texture, or shape in the dynamic video are grouped together to form whole objects. The challenges these systems face are recognizing new objects as they appear in the scene, tracking correctly when objects appear to split due to occlusion, and coping with merging when objects move close together. Approaches range from simple, clustering algorithms [17] to use of the sophisticated probabilistic frameworks [28] we describe in the next chapter. These systems have considerable promise, as they can perform real-time tracking of people and objects without instrumentation. Remaining technical limitations are sensitivity to changes in lighting and the inability to accurately track more than a few people or objects at a time. In addition, there is still the privacy concern that has, thus far, primarily limited the use of these systems to research deployments and public spaces.

**FIGURE 6.5:** Samples images from a real-time vision-based system that can track an object after it is manually identified on the video [28]. © 2003 IEEE.

### 6.2.3   Visual Tags

A number of approaches have been developed to assist vision systems by explicitly placing easily identified features in the environment. This both makes feature recognition faster, as the algorithm can be highly optimized for recognizing these features and also allows the system to operate well in environments lacking the crisp visual discontinuities that algorithms like SIFT detect best.

CyberCode [138] and TRIP [33] are both systems that improve the performance of vision-based location using 2-D barcodes designed to be easily recognized by cameras. Examples of a CyberCode and a TRIP tag are shown in Figure 6.6. Unlike traditional 2-D barcodes, these are not optimized for their data-carrying capacity, but rather were designed to be robustly recognized with inexpensive cameras and to allow the distance from and orientation to the tag to be estimated. Unlike traditional vision features, as the tags are of known size, a 3-D tag's location can be estimated from a single observation. This approach achieves very high accuracy: when observed from 2 m or less, TRIP tags are recognized more than 99% of the time and with median accuracy of 1 cm. Due to the relative simplicity of robust tag recognition, both of these systems can extract tag matches at real-time rates of 15–20 frames per second. These tags are suitable for both helping a mobile camera infer its location using fixed tags and helping fixed cameras track tagged people and objects in their field of view. The obvious disadvantage of using special tags is that all objects must be overtly tagged, and objects with occluded tags are invisible to the system. Köhler et al. have tried to address this shortcoming of assisted vision by using a projector to add the overt features to the scene being observed by the camera [86]. Their TrackSense system uses a mobile camera/projector combination to both create and then measure features to refine an existing location estimate. TrackSense projects a coarse grid pattern onto walls, floors, and ceiling and, using the camera images, infers the location and angle of the visible surfaces and their intersections. By detecting three surfaces (e.g., two walls and a ceiling), the system can estimate the relative 3-D location of the camera to the intersection.

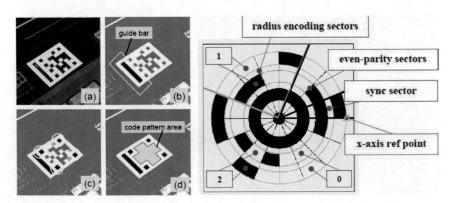

**FIGURE 6.6:** Example of (left) a CyberCode tag and (right) a TRIP tag. Note that both contain information for guiding recognition and data encoding.

From a distance of 2–3 m from a known corner, TrackSense estimates 3-D location with an accuracy of around 20 cm [86]. Because these grid features do not carry identity, this technique needs to be used in conjunction with other location technologies (e.g., an ultrasound room identification beacon and a digital compass) to infer absolute location and orientation.

### 6.2.4   Practical Considerations

Despite the huge amount of deployed camera infrastructure and the undeniable richness of the images captured, the use of vision for automated tracking and location is still uncommon. Much of this is due to the social aversion of being on camera, especially in the home. For example, a study of sensor deployment efficiency by homeowners found reticence to place even mock camera sensors in their residences [11]. Beyond the social issues, vision-based location systems still face technical challenges in practice. Despite many advances, they are still sensitive to changes in lighting, scale, and scene clutter and have trouble tracking multiple things in the same scene simultaneously.

## 6.3   LASER RANGE FINDERS

The laser range finder is a relatively new sensing device that has revolutionized robotic navigation in recent years. The most common type of laser range finder uses a spinning infrared laser to produce an accurate 180° planar "picture" of how far away the closest obstruction is. Figure 6.7 shows a common model of a laser range finder, and Figure 6.8 shows an example of the data they provide. This data is used in the same types of feature-based mapping and localizing algorithms described in the last section. Because of the precision of the readings and the independence from lighting conditions, these systems work very well and allow robots to localize themselves within a centimeter or so in typical indoor environments [46, 114]. The inferences from these same sensors and algorithms have also been the cornerstone of the successful entries in the recent DARPA grand challenges for autonomous vehicles [30]. Due to cost, size, and weight, there has not yet been an egocentric system

**FIGURE 6.7:** A commercial laser range finder that produces a 180° ranging measurements at 75 Hz.

**FIGURE 6.8:** An example of laser range finder data. The range finder on the robot left received the ranging data shown on the right [46].

in which people carry laser range finders to estimate location. Hightower et al. [47] have developed a system that uses wall-mounted laser range finders to track people moving in the environment and merges information from an infrared badge system to add identity.

## 6.4 AUDIBLE SOUND

Two advantages of estimating location with ultrasound are that it propagates slowly enough to make time-of-flight estimates accurate, and it is largely blocked by walls and doors. The audible sounds generated by everyday human activities like talking, eating, and working have the same characteristics. Bian et al. performed a series of experiments in the Georgia Tech Aware Home to determine how accurately inexpensive microphones can detect the location of common activities [14]. Using a total of 16 microphones deployed in four-microphone arrays in the living room and kitchen, they were able to locate where audible activities were performed with median accuracy of approximately 15 cm. Their experiments found that the easiest events to locate involved crisp, sharp sounds such as clicking, snapping, sniffing, or setting thing on hard surfaces. Scott and Dragovich have built on this and explored the use of sharp human-generated noises like hand claps and finger snaps as an input mechanism with both intention and location [150]. (As an example, snapping above or below a light switch on the wall might adjust the dimmer up or down). Their experiments showed similar results to the experiments of Bian et al., with median accuracies of around 10 cm for snapping events.

## 6.5 INTERNET PROTOCOL MEASUREMENT

Nearly all networked devices directly or by proxy use the Internet Protocol or IP for naming and data routing. Techniques have been developed to estimate location from a device's IP address in the event that no other location estimate is available. In practice, these estimates are very coarse, typically at the state or metropolitan granularity. They are also error prone, and connecting a device via VPN, for example, can move its location estimate thousands of miles. One class of techniques is hint-based and looks in various databases for geographic information about the device and its routers. These systems use the domain name system location (DNS LOC) records, the WHOIS information, and even the hostnames themselves [117, 130]. The other approach uses multilateration to estimate location based on end-to-end routing delay [124]. These systems measure the IP latency (typically with an ICMP "ping" request) to the device from a set of landmark machines in known locations. Given these ranging measurements from known coordinates, location estimation is straightforward. One study showed that using a set of Planet Lab hosts [132] as landmarks, simple geometric estimation yields median accuracy of around 200 km, while a constraint-based solution improves median accuracy to around 70 km [80].

## 6.6    MAGNETIC FIELD STRENGTH

Location systems based on magnetic field strength provide a number of our extremes as they offer the highest accuracy, but the lowest coverage and the highest cost. These systems use a fixed transmitter to emit magnetic pulses, and small, body-worn listeners report via radio the field readings they observe. From this data, 3-D location and pitch, yaw, and roll can be estimated with extremely high accuracy at rates up to 120 Hz. The MotionStar system, for example, estimates 3-D locations within 1 mm and angles within 0.1° [5]. By fitting as many as 20 sensors to a person's body at various joints and limbs, these systems allow real-time, high-rate motion capture of human activity. As a result, these systems are one of the primarily means of capturing motion for animated characters in cartoons and video games. The downsides of magnetic-field location systems are the low coverage (they only work when the listeners are within 3 m of the transmitter) and high cost: system starts at around US$50,000.

## 6.7    RADIO FREQUENCY IDENTIFICATION TAGS

Radio frequency identification or RFID tags are battery-free radio devices which broadcast their identity (typically a 96-bit number) when they are remotely energized by an RFID reader. Tags and readers come in many varieties, but are most commonly designed for two different uses: short range, with typical operating ranges of a centimeter or so, which work by inductive coupling; and long range, which use electromagnetic capture at UHF radio frequencies and usually operate at ranges up to 6 m.[1] Both types of tags are small, inexpensive, durable, and as they have no battery to be changed or charged, they can be easily embedded in everyday objects and even be molded into

---

[1]There is a class of radio devices referred to as "active RFID tags." These are small radio transponders with batteries that periodically broadcast their ID. These are RFID tags in the same sense that any 802.11 or GSM device uses radio signals and sends its ID when it transmits. Calling these RFID is confusing, as they are not battery free and are at least an order of magnitude larger and more expensive than true RFID tags. We save our discussion of location systems built on "active RFID" for the next subsection.

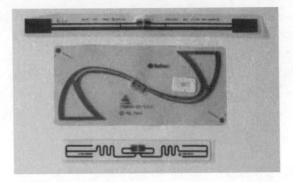

**FIGURE 6.9:** Examples of long-range UHF RFID tags. For context, the center tag is 11 × 5 cm [64]. © 2004 IEEE.

a case without physical access. Figure 6.9 shows three examples of long-range UHF tags. For more information on the theory, technology, and application of RFID, please see the companion lecture *RFID Explained* in this lecture series [170].

One of the first systems to use RFID for long-range location was the Georgia Tech Aware Home, which uses UHF readers placed under floor mats and carpets to track the movement of objects and people around the home [3]. Rather than blanketing the space with readers, they were placed, as shown in Figure 6.10, in strategic locations to measure the movement of tags between rooms. The high reliability with which the readers detect tags makes this approach effective as a room-level location system for tagged objects and people.

Hähnel et al. reverse this approach and use stationary RFID tags to help a mobile robot locate itself using on-board RFID readers [64]. Their system uses a pair of orthogonally mounted

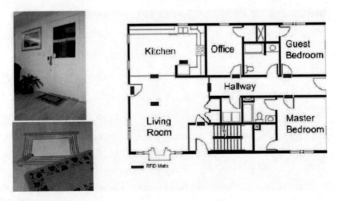

**FIGURE 6.10:** This map shows a placement of RFID readers in the Aware Home that accurately track the transition of tags between rooms [3].

**FIGURE 6.11:** The two white RFID readers can guide this robot with 2 m accuracy without the assistance of the laser range finder [64]. © 2004 IEEE.

915-MHz RFID readers (shown in Figure 6.11) as an alternative to the robot's laser range finder for localization. In their deployment, 100 tags were uniformly spaced through a 1,000-m² office, and the robot used the laser range finder to build a map and estimate the location of the tags. Then, using the RFID tags alone, the robot was able to locate itself with 2 m of median accuracy.

Finally, the Ferret system uses a mobile reader to estimate the location of (possibly) mobile tags [103]. By performing an offline analysis of observations of tags by a mobile reader with known location and orientation, they show that tag location can be estimated within 10 cm, given observations from a variety of angles. Such a system could be used to maintain an offline database of the location of tagged objects if a reader were built into machines like vacuum cleaners and mail carts that are navigated through a space on a regular basis.

As RFID readers cost orders of magnitude more than RFID tags, the choice of mobile tags vs. mobile readers will be affected by desired coverage and the number of objects to be instrumented. The reader antennas used by Hähnel et al. and the Ferret system are too large to be incorporated into a phone or PDA form factor, and miniaturizing long-range readers may not be feasible due the relationship between range and antenna size. The larger surface of a notebook computer screen, on the other hand, could likely accommodate a long-range RFID antenna.

## 6.8    RADIO FROM FM TO ULTRAWIDE BAND

In addition to the GPS, 802.11, and cell-phone-based radio systems, we have presented thus far, a wide variety of other radio-based location systems have been developed. The first two we discuss are both commercially available and are the only two time-of-flight systems in this subsection. The first, UbiSense [165], is an indoor system that uses mobile ultrawide band (UWB) tags (shown in Figure 6.12) that beacon to fixed UWB listeners. The range of the tags is tens of meters, and because both

**FIGURE 6.12:** A UbiSense compact ultrawide-band tag (25 g).

time difference of arrival and angle of arrival are measured, readings are only needed by two listeners to estimate a 3-D location. Like most other mobile-beacon systems, the precise location of the listeners must be known. The UWB spectrum was chosen due to the ease with which multipath reflections can be filtered out by the listeners. In an open environment, the system designers estimate the location estimates to be accurate within 15 cm 95% of the time [161].

The other time-of-arrival system we want to describe is the Rosum Positioning Technology (RPT), a GPS-like time-difference of arrival system that uses signals from terrestrial digital TV towers rather than satellites [142]. RPT makes use of the dozen or more TV stations that are typically broadcasting in an area, and like GPS, only a small number are needed to accurately estimate location. Unlikely GPS, TV signals are typically strong enough to penetrate buildings, allowing RPT to be used indoors as well as outdoors. Unlike GPS satellites, TV towers do not move, meaning the transmitter locations can be cached in the device and do not need to be continually updated or inferred. RPT is initially being deployed for use with the American Television Standards Committee (ATSC) Digital TV signals being broadcast in the continental United States. Other DTV standards like DVB in Europe and ISDB-T in Japan include synchronization signals that could also be used for this kind of location estimation. Testing by Rosum shows accuracy ranging from 3 m outdoors to 10–20 m for indoor urban environments [137]. While not widely available now, digital TV tuning will likely become a common feature on notebook computers, making RPT a potential low cost, high coverage location solution.

One common wireless networking technology we have not discussed yet is Bluetooth. While most Bluetooth devices are mobile, there are stationary devices like printers and base stations that have Bluetooth interfaces, and these can be used for both modeling and fingerprint-based location estimation.[2] One would suspect that the relatively short range of Bluetooth (around 10 m) would

---

[2]While Bluetooth devices can be classified as mobile vs. stationary by observing their mobility over a period of time, there is another clue: Bluetooth advertisements include a product ID which can be used to distinguish mobile from stationary devices.

make it a highly accurate type of radio beacon, especially in comparison to 802.11 and cell phone signals. Unfortunately, Bluetooth does not work well as a location-estimating radio technology. The first issue is one of beacon density: while most mobile devices today have Bluetooth radios, there are far fewer fixed Bluetooth devices. For example, Place Lab's wardrives of Seattle, Washington USA, include hundreds of cell towers, thousands of 802.11 APs, but "virtual no fixed Bluetooth" [95]. This can be attributed to the sparseness of Bluetooth beacons, but also to the approximately 10 s it takes to perform a Bluetooth scan. At this slow scan rate, a scanning device traveling at automobile or even walking speed can miss a Bluetooth beacon as it passes by. LaMarca et al. did perform a controlled experiment that showed Bluetooth beacons, if available, can provide 10 m accuracy using Place Lab's signal modeling algorithm [95]. This accuracy seems surprising given that it is not much better than the 15 m accuracy that signal modeling provides with 802.11, which has a much larger range than Bluetooth. The explanation can be found in an analysis of Bluetooth signal propagation by Madhavapeddy and Tse [105]. They used an Active Bat installation to provide accurate ground truth location, and they mapped the received signal strength and link quality of Bluetooth transmission. Their results show that while Bluetooth devices may have a functional range of around 10 m, packets can be heard up to 25 m away. They also found that, because of the frequency hopping in the protocol, received signal strengths are highly variable once a device is more than a few meters from the transmitter. The low scan rate coupled with the signal strength variability make Bluetooth a poor choice of technology on which to base a location system, independent of the density of devices.

In the RightSpot system, Krumm et al. investigate the feasibility of estimating location by fingerprinting the signal strength of FM radio stations [91]. Using commercial software that took into account tower location, broadcast strength, and topology, they were able to generate a simulated (as opposed to measured) radio map for the 32 FM stations in the Seattle–Bellevue area in Washington, USA. They validated the simulated radio map at sampled locations and found a 95% correlation between the rank-order of station strengths between the simulation and their observations. Because of the enormous area covered by an FM radio station, the median accuracy of their system was only 8 km [175]. Despite the large degree of uncertainty, Krumm et al. felt such location estimates would still be useful for local information such as traffic, weather, and nearby attractions.

SpotOn [69] and LANDMARC [119] are two research location systems that use received signal-strength to track commercial "active RFID" beacons. In the case of SpotOn, the tags send beacons at 916 MHz, while LANDMARC uses 308 MHz. In both cases, the systems achieve accuracy of around 2 m, comparable to that of 802.11 fingerprinting. One interesting innovation in the LANDMARC system is that it uses a dynamic radio map for fingerprinting, rather than a static one. Like RADAR, LANDMARC performs k-mean fingerprinting at a server and assembles a fingerprint from each of the listeners that heard a beacon. In lieu of a static radio map, LANDMARC uses a $1 \times 1$-m grid of tags that are placed on the ceiling in known locations. Each time

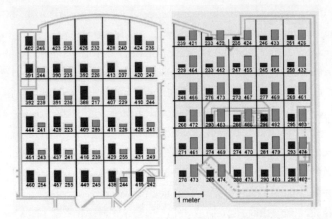

**FIGURE 6.13:** The radio map for two rooms from a PowerLine Positioning deployment [128].

these reference tags send a beacon, the fingerprint for their location in the radio map is updated, allowing LANDMARC to quickly compensate for any changes in the physical environment that alter RF propagation characteristics.

The most unusual of the radio-based techniques is employed by the PowerLine Positioning (PLP) system [128]. In many ways, this system is like other radio fingerprinting systems: it uses fixed beacon that transmit and mobile listeners which employ k-mean fingerprinting with a static radio map to perform location estimation. What makes PLP unusual is that it uses the residential wiring that carries power throughout a home or office as the antenna to broadcast the radio signal. PLP employs a pair of inexpensive commercial tone generators, one at each end of the building, which place a continuous tone (33 and 447 kHz) on the power lines which propagate along the entire electrical system of the home. The unshielded power lines in the home serve as an antenna and transmit the tones so they can be heard wirelessly throughout the home. Because the wiring patterns and the materials that attenuate the signal are likely to be unique, the strength which the signal is received has a high degree of spatial variance. Because home wiring changes seldom happen, the pattern can be expected to have a high degree of temporal stability. Figure 6.13 shows the radio map for two rooms in a PLP deployment where each pair of numbers indicated the observed strength for the two tone frequencies. In the PLP prototype, the signals are received by a custom dongle attached to a mobile computer. In their experiments across a variety of homes and construction materials, PLP achieved median accuracy slightly better than 1 m.

## 6.9    OTHERS STILL TO COME

This chapter presents a roundup of the lesser-used approaches ranging from vision to load sensors to magnetic field distortion. This collection of systems demonstrates that a wide variety of environmental sensors can be used to infer location. This suggests that as mobile devices and smart-space infrastructure incorporate new kinds of sensors and radio (gyros, magnetometers, near-field communication, etc.), it will inspire new location systems with their own particular performance characteristics. Figure 6.14 summarizes the performance of some of the important systems we have presented in this chapter.

| Technology | Smart floors | Vision with fixed cameras | Vision + 2D tags | IP measurement |
|---|---|---|---|---|
| Accuracy | ★★★★ <br> 2D location with 10 – 20 cm accuracy (person may not be IDed) | ★★★★☆ <br> 3D location with 10 – 100 cm depending on camera density | ★★★★ <br> 3D location with 1 cm accuracy when observed from < 2 m away | ★☆☆☆ <br> 2D location with 70-200 km accuracy |
| Coverage | ★★☆☆ <br> In rooms fit with in-floor sensors. Performs poorly with multiple users in room | ★★☆☆ <br> Indoor/outdoor in view of camera | ★★☆☆ <br> Indoor/outdoor in view of camera | ★★★★ <br> worldwide for any networked computer device |
| Infrastructure cost | ★☆☆☆ <br> Cost of retrofitting for load sensors prohibitive | ★★☆☆ <br> Requires networked cameras in each room | ★★☆☆ <br> Requires networked cameras in each room | ★★★★ <br> uses existing IP network |
| Per-client cost | ★★★★ <br> Requires no user instrumentation | ★★★★ <br> Requires no user instrumentation | ★★★☆ <br> Requires 2D tags on objects / people to be tracked | ★★★★ <br> Requires no extra hardware or software |
| Privacy | ★★☆☆ <br> User's location known, but identity often not | ★☆☆☆ <br> Video surveillance among worst intrusions on privacy | ★☆☆☆ <br> Video surveillance among worst intrusions on privacy | ★★★★ <br> No addition information released |
| Well-matched use cases | Assisted living, HVAC/smart home control | Asset and personnel tracking, security / surveillance | Asset and personnel tracking | National/regional information and advertisement delivery |

**FIGURE 6.14:** A summary of the characteristics of other various location estimation technologies.

| Technology | Magnetic field strength | RFID with fixed readers and mobile tags | Ultra-wide band radio TOF/AOA | TDOA from digital TV |
|---|---|---|---|---|
| Accuracy | ★★★★<br>3D location with 1 mm accuracy | ★★★☆<br>Room (or area of room) with high accuracy | ★★★★<br>3D location with 15 cm accuracy | ★★☆☆<br>3D coordinates with 10-20 m median accuracy |
| Coverage | ★☆☆☆<br>User must be within 3 meters of transmitter | ★★☆☆<br>Indoor in room fit with readers | ★★☆☆<br>Indoor in range of UWB listeners | ★★★★<br>indoor/outdoor in range of TV towers |
| Infrastructure cost | ★☆☆☆<br>transmitters cost $50k+ US | ★☆☆☆<br>Long-range RFID readers required for each room | ★★☆☆<br>Requires 2+ UWB listeners | ★★★★<br>piggybacks timing info on existing TV signals |
| Per-client cost | ★☆☆☆<br>Users fit with costly magnetic sensors | ★★★★<br>Each object / person requires inexpensive tag | ★★★☆<br>UWB badge required | ★★★☆<br>antenna and chipset required |
| Privacy | Not a factor for such specialized systems | ★★☆☆<br>Locations tracked by infrastructure. Opt-out by removing tag | ★★☆☆<br>Locations tracked by infrastructure. Opt-out by removing badge | ★★★★<br>Location is estimated passively on the local unit |
| Well-matched use cases | Motion capture for video games and animated media (cartoons) | Asset and personnel tracking, indoor mapping/ navigation/tour guides | Asset and personnel tracking, indoor mapping/ navigation/tour guides | Social networking, assisted living, activity tracking |

**FIGURE 6.14** (*continued*)

CHAPTER 7

# Improving Localization Accuracy

Many of the systems we have presented can perform better in practice than their nominal single-reading accuracy. That is because people and devices are often stationary or moving at slow speed. This means that location systems commonly get dozens, if not hundreds, of opportunities to perform localization, and these results can be combined for greater accuracy.

In this chapter, we review the various ways that have been developed to turn a sequence of environmental observations into a series of location estimates that are more accurate than if each environmental observation was considered by itself. These include simple smoothing approaches that are used to damp out errors and techniques that improve accuracy by constraining reading to the paths and halls of structured environments. We also present probabilistic frameworks that model the motion of the device or person being tracked and can utilize environmental observations produced by multiple different sensors (e.g., GPS, WiFi, GSM beacons) at the same time.

## 7.1    SMOOTHING

In the past few chapters, we have described a multitude of reasons why the location estimates of the system may be noisy and error prone. We have also already described some simple ways wherein smoothing and averaging have been used to lessen the impact of these errors on accuracy. The original RADAR system [8] and a number of follow-on fingerprinting systems [167, 93, 122] employ a $k$-mean clustering algorithm in which the location of the closest $k$ scans in the radio map are averaged together. This is an example of a spatial smoothing technique to damp out reading-to-reading noise and to compensate for sparse areas in a radio map.

Many of the systems we have presented also use a temporal smoothing technique, where all the location estimates in a fixed-sized sliding window are averaged to produce the smoothed estimate. In the case of coordinate-based systems, the average is commonly the geometric mean of the coordinates of the location estimates in the sliding window [67]. In the Cricket system [136], individual location estimates are comprised of a set of beacons and their estimated distances from the listener. To estimate the closest beacon, the Cricket system chooses the beacon with the single lowest mean distance over the window of estimates. Sliding windows have also been applied to systems like SkyLoc [167] that predict symbolic locations (e.g., "Mezzanine Lobby"). This is typically done by choosing the discrete location that appears the most times in the window of estimates.

Choosing a good *k for* spatial smoothing or the window's size for temporal smoothing is important, and a poor choice can overdamp or underdamp the system. If *k* is too small, the system will appear "twitchy," and the location estimates will bounce around from second to second in a distracting way. Conversely, a *k* that is too large can make the system sluggish and non-responsive for mobile users. The "right" *k* is also heavily dependent on technology-specific features like the density of a radio map or concentration of beacons. A common approach for selecting *k* is to do cross validation where a number of different values for *k* are tested on the training data, and the best one is selected. It is common to add a regularizer that penalizes large *k* to prevent overfitting. For 802.11 fingerprinting systems, a *k* of 3–4 is a common choice that balances the two extremes. A recent study by King et al. [83] suggests why: Figure 7.1 shows how the accuracy of their 802.11 fingerprinting system varies with the number of scans a stationary device uses to estimate its location. The knee of this curve is around three or four readings, suggesting that even if arbitrary sluggishness could be tolerated, there is little gain in taking more readings.

A hidden Markov model (HMM) is a more powerful way to smooth results in location systems that estimate discrete or symbolic locations. HMMs are variants of state machines with states and transition probabilities in which the state is unknown, but can be estimated from observations. In location systems, HMMs have been used to choose radio readings in WiFi fingerprinting systems [93], to predict routes from GPS data [108], to predict place names from GPS readings [6], and to estimate whether a device was moving or stationary [89]. Beyond HMMs, a number of other

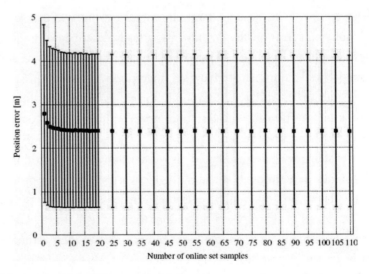

**FIGURE 7.1:** This graph shows how accuracy improves as a stationary device performing 802.11 fingerprinting is permitted more scans to estimate its location [83].

probabilistic techniques are highly effective at smoothing out noisy sensor data; as these techniques also enable temporal tracking and fusion across location technologies, we discuss them in Section 7.3.

## 7.2   SNAPPING

Another common technique for reducing the error in location systems is to constrain coordinate-based location estimates using the natural boundaries in the physical environment. The best examples of this are the GPS-based in-car navigation systems. Because it is extremely likely that drivers confine their navigation to roads, the systems snap the GPS readings to the closest point on the nearest segment of roadway. Assuming the car is indeed on the road, this eliminates any GPS error perpendicular to the road's path and, in effect, halves the GPS error. In practice, this trick improves perceived accuracy far more, as people are better at judging their distance off the road segment than along it.

These techniques have also been applied to indoor location systems in which locations are constrained to a graph that spans the common areas and hallways of an office environment. Figure 7.2 depicts a graph used to represent location estimates in an indoor location system employing infrared and ultrasonic beacons [47]. The main limitation of this approach is that it does not accommodate unconstrained movement in open spaces. Whereas the significance of this issue is likely to be limited in the office space shown in Figure 7.2, it would definitely cause a significant problem

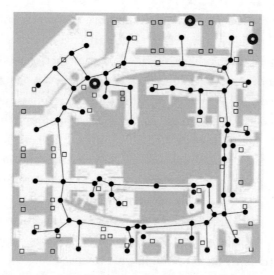

**FIGURE 7.2:** This map shows the graph used to constrain the location estimates in an indoor location system that uses infrared (shown as small squares) and ultrasonic (shown as circles) beacons to locate the user [47].

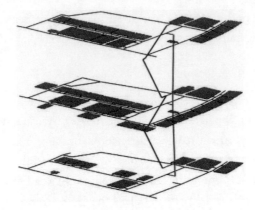

**FIGURE 7.3:** A schematics of a multistory building using a mixed representation where hallways, stair-cases, and elevators are modeled as graph edges and open spaces are modeled as bounded free space [44].

in large open spaces. A solution to this challenge has been proposed by Ferris et al. in the form of a mixed space representation that employs both graphs for constrained spaces like paths and halls as well as bounded open spaces in which motion is unconstrained [44]. Figure 7.3 shows a sample map of a building in this mixed representation.

## 7.3    FUSING AND TRACKING

*Fusing* and *tracking* are the other two key ways to improve the accuracy of location estimation. Fusion, in the context of a location system, refers to the effective use of two or more different types of sensor observations to determine the position of the same person or device. Tracking improves position estimation by modeling the motion of the device or person being tracked and includes the capability to continue producing a stream of updated location estimates even during periods of unavailability of new sensor data. Both of these are difficult capabilities to incorporate in a principled manner into simple modeling techniques such as a centroid, *k-mean* clustering or window-based smoothing. For example, there is no simple way to compute the centroid of a GPS reading, a cell tower observation, and two ultrasonic ranging measurements. Similarly, in the absence of new readings, it is unclear how we use a sliding window of location estimates to track a user. Fortunately, there are probabilistic reasoning techniques that are well suited for these tasks, the most popular of which are *Bayes filters* [145]. Broadly speaking, Bayes filters perform state estimation on a dynamic system when there is uncertainty due to noise in the evidence being collected. This fits the task of location estimation from sensor readings, as the state we are trying to estimate is easy to define, and the system typically has a continuous but often noisy stream of environmental observations. As a result, Bayes filters have been used in dozens of research and commercial location systems.

Other machine learning techniques have been used for location estimation including fuzzy logic [146], neural networks [10], and support vector machines [148], but because of their overwhelming popularity and applicability, we focus our discussion on formulations of Bayes filters and their uses and trade-offs in location tracking. For further reading, we would direct the reader to the excellent tutorial of Fox et al. on the use of Bayes filters for location estimation [47].

Bayes filters are derived from Bayes' rule which says how to revise probability distributions in light of new evidence. Assuming we have a prior estimate of the likely state of variable $X$ and we are given new evidence $E$, Bayes' rule says that the new probability distribution for $X$ is

$$\text{Bayes' rule:} \quad P(X \mid E) = \frac{P(E \mid X)P(X)}{P(E)}$$

$P(X)$ is the prior probability or what we thought about the state of $X$ before the evidence, and the factor $P(E|X)/P(E)$ captures the effect that the evidence has on our belief about $X$. The resulting probability distribution $P(X|E)$ is known as the posterior probability. The purpose of a Bayes filters is to track the belief about $X$ over a series of observations. That is to say, they compute $P(X_t|E_1,E_2,E_3,\ldots,E_t)$. (In location terms, Bayes filters are used to estimate the chance of a device being in a given location having seen the sequence of sensor readings $E_1,E_2,E_3,\ldots,E_t$.) To keep a recursive expansion of Bayes' rule from blowing up exponentially, Bayes filters assume that the system has the *Markov property*, meaning that the future state of $X$ depends only on its current state and not on any past states. This assumption of the independence of the current state and how it was reached allows Bayes filters to define beliefs about $X_t$ inductively in terms of $X_{t-1}$:

$$\text{Bayes filter:} \quad P(X_t) = nP(E_t \mid X_t)\int P(X_t \mid X_{t-1})P(X_{t-1})\mathrm{d}X_{t-1}$$

In this equation, the conditional probability $P(X_t \mid X_{t-1})$ describes the dynamics of the system. In a location system, this is known as the *motion model*, and it quantifies the chance that, since the last reading, the person or device has moved from location $X_{t-1}$ to location $X_t$. As an example from a real system: Hightower et al. represent their indoor motion model as a Gaussian with a mean of 0.5 m/s clipped at 10.2 m/s (the maximum human speed on foot) [67]. The Opportunity Knocks project used outdoor motion models that predict movement based on the estimated mode of transportation (walking, riding bus, etc) [129].

The other conditional probability $P(E_t|X_t)$ describes the *sensor model*, which captures the likelihood of observing sensor reading $E_t$ at location $X_t$. GPS is commonly modeled using a Gaussian with a mean error of 8 m (or better depending on the GPS variant). Muller et al. use a quadratic likelihood function to model Doppler shift in ultrasound [115]. As we saw in Chapter 4, variations in radio signal strength are Gaussian, thus Gaussians of various spreads are commonly used to represent the sensor models for wide-area 802.11 and GSM localization [95]. Recently, more accurate sensor models for 802.11 and GSM have been developed that are based on Gaussian Processes [44].

The last factor in the equation $n$ serves to normalize the posterior distribution to keep the probabilities summing to 1. In terms of tracking the location of a device, the equation for the Bayes filter can be interpreted as: To figure out the chance that the device is currently at location $\alpha$, consider (based on where the device was last thought to be) the likelihood that it would move to location $\alpha$ and then observe sensor reading $E$. Before considering how to represent these models computationally, consider a few characteristics of Bayes filters implied by this formula.

### 7.3.1   Sensor Fusion

Once a technology (GPS, ultrasound TOF, etc.) has been modeled as a function describing the likelihood of receiving a reading at a given location, it can be utilized in conjunction with any other technologies in a Bayes filter. We will see in the coming sections that virtually all the technologies we have described have been modeled probabilistically, and this has enabled both wide-area systems like Place Lab [95] (which can fuse GPS, 802.11, Bluetooth, and GSM reading) and indoor systems like the Enterprise Indoor Location System [71], which includes sensor models for IR badges, active and passive RFID, laser range finders, 802.11 fingerprints, and GPS.

### 7.3.2   Tracking Ability

The state variable $X$ may be simple location, but can also include dimensions such as velocity and heading. If velocity and heading can be accurately estimated by the sensor observations and are utilized by the motion model, tracking behavior emerges naturally from a Bayes filter, as in the absence of any new sensor readings, the sole conditional probability operating is the motion model.

### 7.3.3   Tolerance to Noise

The explicit modeling of both sensing and motion make Bayes filters good at tolerating noisy sensor data. An errant GPS reading may indicate, for example, that the user has moved 100 km north since the reading 1 s before. Assuming a decent model of GPS error and human motion, the Bayes filter would determine that the most likely places to hear that reading were virtually impossible to reach in 1 s and would instead conclude that the user had probably moved a small distance and had seen an unlikely, but not uncommonly, inaccurate GPS reading.

We now describe both parametric and sample-based implementations of Bayes filters and how they have been optimized to be run in real-time on resource-challenges hand-held devices.

### 7.3.4   Kalman Filters

Kalman filters are Bayes filters which represent their belief as a first and second moment, which very closely models a unimodal Gaussian distribution [109]. Kalman filters have long been a mainstay of the aerospace field and were used, for example, to filter the Doppler radar data that was used to guide the descent of the Apollo 11 lunar module. Kalman filters have two large advantages. First, if the uncertainty in a system is Gaussian, and the dynamics of the system are linear, a Kalman filter has been shown to be optimal. Second, as they represent belief as a relatively small vector of numbers, Kalman filters require little memory and use simple matrix operations to update. Due to these advantages, Kalman filters have been used in airplane, car, and handheld location systems.

The Kalman filter's unimodal belief representation does somewhat limit its uses. Consider tracking a car that is approaching a fork in the road. If the GPS data were noisy, a system might want to maintain two different hypotheses: that the car might have traveled down the left fork or the right fork, something not possible with a basic Kalman filter. To address this drawback, variants of the basic Kalman filters have been developed. Belief has been modeled as a combination of multiple Gaussians to allow more than one hypotheses to be maintained at the same time [101]. A number of research systems have taken this to an extreme and overlay a grid onto the coordinate space and maintain a separate hypothesis for each grid square. The Predestination system [90], for example, employs a 1 km by 1 km grid to maintain hypotheses about a user's driving destination on a county road map (shown in Figure 7.4). Based on what is known about that square kilometers roads, zoning, and population, where the user has driven in the past, and the partial route driven

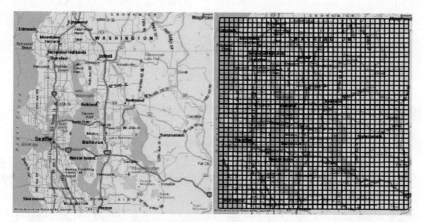

**FIGURE 7.4:** Krumm and Horvitz divide the Seattle Metro into 1 km square grid squares each with its own probability distribution [90].

so far, a Kalman filter estimates the likelihood that each grid square is the user's ultimate destination. Similar gridding techniques have been used for indoor location system, typically with grid squares of around a meter. This approach creates a trade-off between the number of simultaneous hypotheses to be maintained and the space and computational overhead of the filter. In the limit, if the area is too large or the grid is too dense, it becomes impossible to update the filter in real-time. Fortunately, there is a Monte Carlo sampling method for Bayesian filtering we will present next which largely addresses this problem.

### 7.3.5 Particle Filters

Particle filters are a Monte Carlo method for representing posterior distributions with a collection of individual samples or particles [37]. Each of these particles is a fully formed hypothesis; in the case of a location system, a particle might represent the hypothesis that the device is currently at 43°39′39.95″N, 79°23′44.36″W. In a particle filter, the density of particles reflects likelihood, and given a sufficient number of particles, an arbitrary probability distribution can be represented. Figure 7.5 shows a visualization of a particle filter performing indoor location estimation [67]. In each image, the true location of the user is indicated by the path originating from the person icon, and the small dots represent the hypotheses of the particle filter. The first image on the left shows the particles uniformly spread about the office, as the system is in its initial state and has received no readings. The middle image shows the updated state of the particle filter after receiving an infrared reading from the indicated beacon. Note that while the particles are still spread widely, they have largely converged on the space around the beacon. The final image shows the state after receiving an ultrasonic ranging reading. The circle around the ultrasound beacon, in this third image, denotes the most likely place to have observed this reading. Note that the readings in the right image are largely confined to the right side of the beacon despite it having a symmetric sensor model. This is due to the motion model which predicts that it is more likely for particles to have moved from

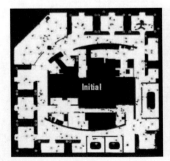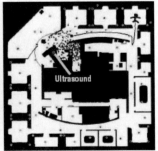

**FIGURE 7.5:** A particle filter tracking a person using both ultrasound and infrared readings [67].

their old positions to the right side of this beacon than the left. We can see the effect that this fusion and motion tracking have on accuracy in Figure 7.6. This graph compares the accuracy of the particle filter to a number of the simple non-probabilistic techniques we have described. Despite getting the same sensor readings, the particle filter is far more accurate, with a median error of around 1 vs. 4 m for the simpler techniques [67, 79].

Figure 7.7 is another visualization of particle filter updates, this time doing outdoor localization [62]. In this series of images, the particles are again small dots, and the true location of the vehicle is represented as an "x." In this experiment, the only sensor data the system receives are indications of left and right turns by the car and the wheel rotations between turns. The system does, however, know the road map and can assume the car is staying on the road. As a result, the initial state of the particle filter (shown left) starts with the particles uniformly spread on the road segments. The middle image shows that after two turns by the driver, the particle filter has converged on a half-dozen particle clusters indicating likely locations. (This center image is a good example of a belief that could not be represented with a unimodal Kalman filter.) By the fifth turn, the right image shows that the particles in the filter have converged into a single cloud that is tightly grouped around the true location of the car.

A number of algorithms and subsequent improvements have been developed for iterating the state of a particle filter on receiving new evidence. The most basic is the sequential importance sampling, or SIR, algorithm, and it works by first drawing a set of sample particles from the full set based on the particle's weight or importance. (Particles are given uniform weight when the filter is initialized.) Each sample in the selected set is then updated (e.g., moved), based on the motion model, and its weight is changed to reflect the likelihood of having moved as it did and then ob-

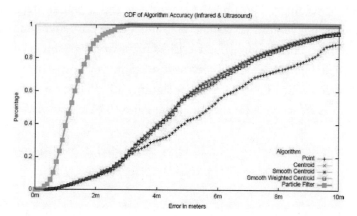

**FIGURE 7.6:** Accuracy of a location system using particle filter fusion of infrared beacon and ultrasonic ranging observations [67].

**FIGURE 7.7:** These three images show a particle filter tracking the location of a car using turn information alone. With each successive turn the system become more confident of the street the car is on. [62]. © 2002 IEEE.

served the new evidence. The new set of weighted particles becomes the posterior distribution from which samples are drawn on the next iteration. SIR suffers from tendency to degenerate to only a few high-weight particles; thus, a resampling step is commonly performed after each iteration as well [55]. Another common improvement is to adjust the number of particles in the system to balance the efficiency and accuracy of the filter [45]. These changes have allowed real-time particle filter-based location systems to be run on platforms as small as PDAs [67] and mobile phones [95].

## 7.4  SUMMARY

In this chapter, we present a variety of ways to improve the accuracy of a location system by interpolating across sensor readings. The simplest smoothing and averaging techniques provide only modest improvements, but require very small amounts of memory and computation and can be implemented on even the most resource impoverished platforms. Probabilistic frameworks like Kalman filters and particle filters offer the ability to fuse readings across sensor technologies and incorporate motion models which improve accuracy significantly. When carefully implemented, these probabilistic techniques are suitable for mobile phones and PDA platforms.

· · · ·

CHAPTER 8

# Location-Based Applications and Services

In this chapter, we examine a wide variety of application domains, ranging from the life-and-death context of emergency response to serendipitous social meet-ups, in which location-aware devices are used. We describe both emerging and well-established applications that make use of nearly all of the location technologies we have described thus far. It will be apparent, however, that GPS is used in a far wider variety of applications than any other location technology. This is in large part due to the high coverage GPS offers outdoors and the wide variety of GPS form factors that are available. We conclude the chapter with a brief overview of tools for building location-aware applications.

## 8.1 NAVIGATION AND WAY-FINDING

The most wide-spread application of location technology is outdoor navigation and myriad of products are available to help people figure out where they are and how to get where they want to go. Nearly all of these systems use GPS for their location estimation, as the accuracy and coverage of GPS is well suited for most outdoor navigation tasks. Specific devices have been designed with the requirements of urban pedestrians, hikers, automobiles, airplanes, and boats in mind. The most impoverished systems present the user with only their latitude and longitude, but most incorporate the capability to show users where they are on a map (as shown in Figure 8.1). New maps can typically be added using PC software or a removable flash memory, although some newer offerings can use wireless LAN or WAN capability to automatically download maps upon entering a new area. Another common feature is for navigation systems to make use of street topology to provide turn-by-turn driving directions from the current location to a destination address. While traditionally found only in navigation systems for luxury and rental cars, this feature is increasingly common in less expensive mobile units such as the popular Nüvi shown in the right side of Figure 8.1. The most recent feature to appear in navigation devices is the ability to continually download real-time traffic data, which is used by the device to choose routes that avoid congestion.

**FIGURE 8.1:** Two GPS navigation devices. The left image shows a Garmin GPSMAP 378, a unit intended for marine use. The right image shows the Garmin Nüvi, a popular car navigation system that provides turn-by-turn driving directions.

The use of location technology for indoor navigation is far less common, and indoor navigation systems for people (such as the system shown in Figure 6.3) are still found only in the research and prototype stage. This is mainly because, for most indoor environments, indoor navigation is not a terribly motivating application for those that spend the most time and invest the most money in a building, as they are unlikely to need navigation assistance themselves. Nevertheless, there are indoor venues such as malls and museums, and conditions such as cognitive and visual impairment which are well suited for indoor navigation. As a final thought on navigation, we introduced in Chapter 6 technologies that have been used by robots to perform localization and navigation. While uncommon, autonomous robots are used to deliver supplies and materials in commercial and industrial settings. These robotic systems typically employ a self-mapping capability to learn the routes and obstructions in the environments in which they are deployed [114].

## 8.2   ASSET TRACKING

Printed bar codes and RFID tags are the technologies most commonly associated with asset tracking. These technologies, as traditionally used, are only aware of the location of a package, shipping container, or vehicle when it is in range of a barcode or RFID reader, and these are typically installed in the loading docks, shipping centers, and staging areas through which the assets pass. While less common, location tracking technologies are sometimes used in applications that require real-time or continuous tracking of assets.

Commercial truck and bus fleets are often tracked in real time using GPS. The information is typically used to improve dispatch and routing efficiency and provide real-time ETA information (two examples are shown in Figure 8.2). Vehicle tracking via GPS is also used to locate and recover stolen vehicles and heavy construction equipment.

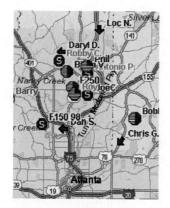 

**FIGURE 8.2:** Two examples of real-time fleet management maps. The left map shows an example of GPS North America's interface for tracking a company's vehicle fleet. The map on the right shows the real-time location of Helsinki's public busses.

Indoor employee tracking is used in time-critical, human-intensive environments like factories, hospitals, and assisted living facilities. Tracking in these environments is primarily done using badge systems like those described in Chapter 3. Figure 8.3 shows three commercial "asset tags" designed to be mounted on mobile equipment. Information from these tags can be used to find lost or missing equipment and to tag highly mobile equipment like cardiac arrest crash carts and gurneys to allow staff to find one quickly.

## 8.3 EMERGENCY RESPONSE

One of the highest value applications of location technology is to aid in assisting individuals who may be lost, hurt, or in danger. E911 is short for "Enhanced 911" and refers to a U.S. Government mandate to provide emergency response personnel with location information for distress calls.

**FIGURE 8.3:** Three asset tags capable of providing real-time location information. Shown left is an infrared tag made by Versus, shown middle is an infrared tag made by Visonic, and shown right is a 802.11 beacon tag made by Ekahau.

The specific accuracy requirements vary for mobile service providers, but the more stringent of the "Phase II" guidelines require that a cell phone handset be localized within 50 m 66% of the time [1]. Similar requirements exist in Europe with E999 and E112. A wide variety of technologies have been used to meet the E911 guidelines including time of flight of mobile phone signal as well as GPS [178]. Despite huge investments, cell phone carriers have been challenged to meet the requirements, especially in rural areas where customers may be as far as 30 km from the nearest cell tower. The situation is even more challenging for providers of Voice Over IP or VOIP phone service as a VOIP service provider may know little or nothing about the networking infrastructure being used to carry the voice data.

There are subscription-based emergency response systems similar to E911 available for both cell phone and cars. The OnStar system, for example, is available as an option on many new cars and uses GPS for location estimation and the cellular phone network for communication [120]. In addition to concierge services, like finding restaurants and hotels, OnStar includes a one-button emergency request that relays the car's location and request for help to a nearby emergency service provider.

## 8.4    GEOFENCING

Location systems are used in a number of niche applications for tracking individuals and signaling when they enter or exit predefined areas of interest; we mention a few of these "geofencing" applications here. Worried parents can choose from a variety of services that track the location of their child's cell phone in real time. The LG Migo (shown in Figure 8.4) is an example of a CDMA phone that Verizon customers can track using the "Chaperone" service. In addition to providing real-time location queries of the handset, Chaperone allows parents to define important locations like home or school and receive SMS text alerts when the handset reaches one of these places. Given the accuracy claims of 50–150 m, Chaperone likely uses the same location estimation as their E911 service.

Location systems have recently made an entrée into the "judicial monitoring" market of products used to provide a prison alternative to those who have run afoul of the law. Older bracelets that enforced a house arrest policy simply measured their proximity from a central base station and would sound an alarm if the monitored individual strayed too far. The newer models (an example of which is shown in Figure 8.5) use both GPS- and cell tower-based location to provide real-time indoor and outdoor coverage. Actual location information provides more flexibility than simple proximity allowing offenders to potentially travel from home to work and back at scheduled times and allowing the definition of exclusion zones that should not be visited.

**FIGURE 8.4:** The LG Migo: A CDMA cell phone, designed primarily for children. With a Migo, Verizon subscribers can request their children's current location and sign up for SMS alerts when the phone reaches predefined areas (home, work, soccer field, etc).

Worldwide location tracking has even been made accessible for pets. The Global Pet Finder system (shown in Figure 8.6) features a GPS receiver and cellular transceiver packed into a 200-g dog collar that can alert their owner if they stray from one of a set of defined zones.

Location technology has recently been proposed as a way to help secure 802.11 networks. Without taking extreme steps like using radio-absobring building materials, even the most careful network deployments allow an 802.11 network to be accessed well beyond the walls of the building. Unfortunately, both WEP and 802.1x, the MAC layer authentication solutions intended to secure 802.11, have well-publicized weaknesses, and both are known to be easily compromised. Garg and

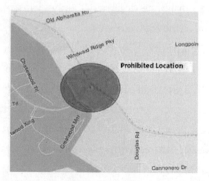

**FIGURE 8.5:** OmniLink's "Offender Tracking Technology" bracelet uses GPS and cell-tower location for indoor and outdoor location estimation. The ankle bracelet is shown left, and a snippet of their profile creation tool is shown right.

**FIGURE 8.6:** The 200-g Global Pet Finder collar uses GPS to locate and the cellular network to alert owners of a pet's location when it has wandered.

others have proposed that server-side location esimation (where APs estimate the client's location) can be used as an additional authentication mechanism [51]. This use of geofencing allows network administrators to define a boundary beyond which 802.11 clients will be ignored. Given a good algorithm and radio map, this would provide access control with 1–3 m accuracy with respect to this boundary.

## 8.5 LOCATION-BASED CONTENT AND SEARCH

A wide variety of devices and applications enable users to both retrieve content based on location and produce content tagged with location. Many popular Internet services such as online yellow pages retrieve content based on a manually entered location. What distinguished the systems in this section is that they retrieve or tag content automatically based on the user's location at the time.

Because a tourist or a visitor is a likely person to want information about their current location, location-aware tourist guides were among the first such systems. Two early systems developed in the 1990s were Cyberguide and the Lancaster GUIDE system. Cyberguide was a PDA-based information appliance designed to give information for visitors to the Georgia Tech campus [2]. An outdoor, campus-wide version of the system used GPS, while a version that guided visitors through the computer science labs used infrared beacons for location estimation. The GUIDE system provided a location-aware walking tour of the city of Lancaster using a tablet computer [26]. The GUIDE system used coarser location than Cyberguide and based location estimates on the ID of the 802.11 access point the tablet was associated with. Many such systems have been developed since and commonly use maps to show both the user's location and nearby points of interest. These systems share features with car navigation systems that include point-of-interest databases and location-aware city guides, such as those designed by Vindigo [168] for PDAs and other mobile devices. The tourist systems distinguish themselves by offering rich, multimedia content on the exhibits and sites that the users may drill down into from the map-based interfaces. Figure 8.7 shows a screenshot of Cyberguide and an example of a commercial device designed for tourist assistance.

The act of tagging content with a relevant location is called "geocoding," and applications have been developed to help tourists geocode the photos they take on a trip. These systems typically

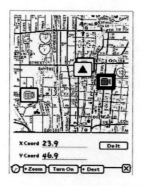 

**FIGURE 8.7:** Shown left is a screenshot of the Cyberguide system from 1996. Shown right is the Evadeos device designed and sold by France's National Geographical Institute. The Evadeos features GPS, a database with 1.7 million points of interest and a facility to add new content via radio or flash memory.

use a GPS logging device to keep track of where they travel throughout the day. A web or PC application later correlates the timestamps from the user's digital photos with the location log to figure out where the photo was taken. Sony sells a GPS device, shown in Figure 8.8, designed specifically for this use case: it has long battery life, no screen, and is small enough to attach to a camera's strap to ensure it gets taken along. The location information allows people to share, search, and browse their photos using a map interface rather than by name or chronology [164]. Figure 8.9 show an example of an application that presents location-tagged photos.

One type of location-based content system most of us unknowingly interact with on a daily basis is the Internet ad server. These systems are responsible for serving billions of ad images to browsers as users surf the web. While ad selection is primarily based on the page the ad is embedded

**FIGURE 8.8:** The Sony GPSCS1KA shown above weighs only 55 g (without battery) and is not much larger than the AA battery it runs on. This unit is designed to create a GPS log that allows digital photos be tagged by location based on their time stamp.

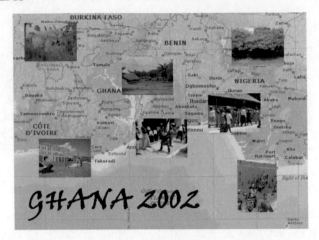

**FIGURE 8.9:** As an alternative to name or time-based image browsing, systems like the World-Wide Media Exchange provide location-based image interfaces.

in and the browser's history, increasingly, the user's location is factored in as well. These systems use one of the IP-based location techniques described in Chapter 6 to estimate where the user is physically located. While only accurate at the county or metropolitan level, the estimate does let the ad provider alert the user that "Mortgage rates have dropped in the Pittsburgh area again!"

There are a handful of applications which involve both mobile authoring and mobile consumption of location-enhanced content [20]. On the serendipitous side, the E-Graffiti system allows users to create a text message and (virtually) leave it in that location as "graffiti" for others to later see. Via their mobile devices, users of the system can look for and read nearby digital graffiti left by others [18]. The goal of the project is to allow people, like those leaving real graffiti, to add small

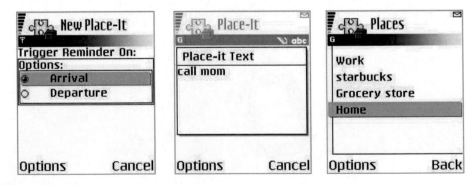

**FIGURE 8.10:** This sequence of phone screens illustrate how users of the Place-Its system author new location-based reminders, in this case, to "Call mom when I get home."

bits of information over time to add detail, character, and information to the environment. Like many of the other systems in this chapter, E-Graffiti uses 802.11 for networking and GPS for location estimation. *Place Its* is a similar system that allows users to create location-tagged reminders, which will be presented when they next visit the target location [156]. With Place Its, a user could, for example, create a reminder to call Mom the next time they arrive at home (Figure 8.10 shows a sequence of Place Its screens authoring this reminder). Place Its is phone based, and reminders are authored, and notifications are served on a smart phone. When a new symbolic location (like "Home") is defined in Place Its, a GSM fingerprint is collected and associated with this place. In the background, Place Its monitors the GSM radio, and if it deems the GSM environment sufficiently close enough to any of the pending reminders' fingerprint, it triggers a notification.

## 8.6    SOCIAL NETWORKING

Location technology is being explored as an enabling technology for social networking as a way to help people coordinate interactions, understand social patterns, and meet new people. Consolvo et al. explored people's attitudes about location disclosure among friends, workmates, and family [9, 22]. Not surprisingly, they found users to have privacy concerns, especially about work colleagues and supervisors having access to their location. Interestingly, they identified high user value in the exchange of location information between friends and family. Much of the value stemmed from the shared context people have that allows location to serve as a proxy for activity, role, intention, etc.

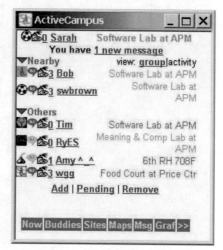

**FIGURE 8.11:** Active Campus includes both a map-based social awareness tool (shown left) in addition to the more traditional buddy-list tool enhanced with location information (shown right) [59].

As part of their MMode Service, AT&T offers a "Friend Finder" feature that allows an AT&T cell phone owner to opt-in (on a per-friend basis) to allow their phone to be located. The service does not allow general location queries or tracking, but rather, like Place Its, serves a notification when a friend is nearby (typically with a hundred meters or so).

In Chapter 3, we introduced Active Campus, the system for which the original wide-area 802.11 location system was developed. While Active Campus included classroom-specific features for asking questions and commenting on course material, it also contains generic social networking tools [59]. Figure 8.11 shows both the map-based and buddy-list style friend-finder in Active Campus.

Dodgeball is a location-based social networking application aimed at people going on the town [35]. Using their cell phones and SMS, Dodgeball allows people to advertise which club or bar they are at and see who else is currently at various venues.

## 8.7  HEALTH AND WELLNESS

GPS is commonly used to track outdoor fitness activities. A popular Google Map mashups, for example, is GMap Pedometer that lets people visualize and share their walking route based on a GPS trace [54] (Figure 8.12 shows a map from this service).

A number of fitness-specific GPS devices have been developed for this niche. Unlike traditional GPS units, these typically have few or no navigation features and are designed to be small,

**FIGURE 8.12:** An example of a map from GMaps Pedometer, a Google Maps mashup that maps walking routes based on uploaded GPS traces.

**FIGURE 8.13:** The Garmin Forerunner 305 is a wrist-worn GPS unit designed for tracking outdoor fitness activities.

rugged, and simply track how quickly and where the user goes. Models such as the Garmin Fore-runner 305 (shown in Figure 8.13) are designed for aerobic activities like running and cycling and incorporate a wireless heart monitor to allow exertion level to be correlated with speed, grade, and location. Once connected to a PC, the activity data is downloaded, and maps, graphs, and reports, such as those shown in Figure 8.14 can be created. While not as accurate, and, strictly speaking, not based on location, Sohn et al. have demonstrated that the change in observed signal strength of GSM cell towers can be used to predict a user's walking, cycling, or driving pace [157]. One of their motivating applications for this capability is a fitness tracking application for GSM phones that does not require extra hardware.

In an application that is part health and wellness and part emergency response, location systems are used to detect wandering in individuals with Alzheimer and other conditions involving

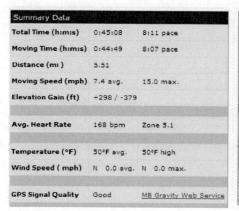

**FIGURE 8.14:** Part of a report from motionbased.com, a web service that lets athletes upload GPS traces and view their activities and progress.

**FIGURE 8.15:** OmniLink's Alzheimer Tracking Solution uses both GPS and cell-tower location for indoor and outdoor location estimation.

dementia. Often in a pendant, bracelet, or anklet form factor, these systems can detect that an individual has wandered and can help locate and assist them. An example of a device designed for individuals with Alzheimer is shown in Figure 8.15. It might not surprise the reader to learn that this device is sold by the same manufacturer as the tracking device shown in Figure 8.5.

## 8.8 GAMING AND ENTERTAINMENT

Location-based games have been proposed as a way to merge the fun and captivation of electronic gaming with the health benefits of being physically active. Especially in the light of the recent teen obesity epidemic, this is seen as a potentially high-value application domain for mobile computing.

The most popular location-based game is *geocaching*. While there are many variants, basic geocaching is an outdoor treasure hunt in which the participants use GPS to locate waterproof containers called "geocaches," which typically contain a logbook to sign or a collection of small plastic toys or trinkets to take as a prize. There are many websites (geocaching.com being the largest) that document the location, contents, and difficulty of literally hundreds of thousands of geocaches in over 100 countries. Figure 8.16 shows photos of two geocaches.

While there have yet to be any commercially successful multiplayer games that utilize location systems, a number of research deployments have explored the challenges in delivering a consistent, enjoyable experience. In one of the first such experimental games, *Can You See Me Now*, and its follow on, *Uncle Roy All Around You*, online players chased each other through a combination of virtual and physical urban spaces [12]. These systems relied on GPS for location information, and this proved to be one of the major sources of poor experiences. The problems were not due to the average GPS error, which the game was designed to accommodate. Rather, issues arose during the periods of extreme inaccuracy when the GPS error rose to 100 m or more. *Can You See Me Now* relied on 802.11 for network connectivity, and disconnections were also an issue despite careful planning. As a result, the follow on system used GPRS rather than 802.11. Barkhuus et al. took

**FIGURE 8.16:** Two examples of geocaches, the left in Germany, the right in the UK.

a different approach in their location-based game called *Treasure* in which a collection of players run around collecting virtual coins, returning them to their virtual treasure chest [9]. Like *Can You See Me Now*, *Treasure* uses GPS for location and 802.11 for communication. Rather than trying to ensure perfect 802.11 coverage, *Treasure* incorporates network disconnection into the gameplay, and when disconnected, players are immune from having their coins stolen. This adds a strategy to the game where players actively seek out the voids in network coverage and use them for safe travel. In *Can You See Me Now* and *Treasure*, players carry a PDA that they interact with as they move around in the physical world. An alternative experience is explored by the *Human Pacman* game in which players take the role of pacmen gobbling treasure or ghosts chasing the pacmen [25]. This system (again using GPS for location and 802.11 for communication) uses an augmented reality overlay to

**FIGURE 8.17:** Two location-based game interfaces. The PDA display for *Treasure* is shown left [9]; the augmented-reality display for *Human Pacman* is shown right [25].

convey to users where in the real world the virtual treasures lie. Screenshots of *Treasure* and *Human Pacman* are shown in Figure 8.17.

## 8.9    CHALLENGES

While many location-enhanced applications and services have been demonstrated, only a handful has reached wide-scale adoption. This is partly due to the newness of location-enhanced computing, but is largely caused by a few specific factors. The first is that most of the mobile devices deployed today are not equipped with a location system. Fortunately, there are a number of free and commercial libraries that perform wide-area location estimation without additional hardware. PlaceEngine [131], Navizon [116], and Skyhook Wireless [154] provide real time 802.11 location estimation for a variety of programming environments. The Privacy Observant Location System (POLS) provides GSM location estimation for Windows Mobile devices [134]. Place Lab is an

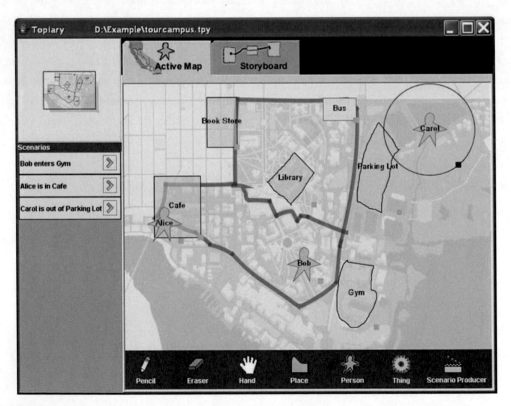

**FIGURE 8.18:** The main interface for Topiary, a rapid prototyping environment for location-enhanced applications [102].

open source Java toolkit that can simultaneously estimate location from GPS, 802.11, Bluetooth, and GSM readings [95].

Another factor is that most location-based applications are tied to a single location system. In recent years, common programming platforms have tried to address this by adding API support for location. JSR-179 is the Java technology specification for the Location API for J2ME [77]. JSR-179 supports a variety of location concepts including coordinate systems, landmarks, location providers, and location estimates. Reference implementations of JSR-179 that wrap generic GPS devices have been developed, and some Java-based location systems like Place Lab provide a JSR-179 interface. Similarly, Microsoft has included a uniform API for location information starting with Windows Mobile 5.

One challenge facing developers is the difficulty of designing location-enhanced interactions and testing them in a variety of conditions. Topiary is a rapid prototyping system designed to alleviate this problem and let developers quickly assemble and experiment with location-enhanced applications [102]. Topiary allows designers to create maps and populate them with people, places, and things. Using rules to define how the application should react based on the location of people and things, Topiary allows rich location-based interactions to be designed, tested, and quickly iterated. A screenshot of the main Topiary workspace is shown in Figure 8.18.

The final factor slowing the spread of location-enhanced applications and services is privacy. Location privacy is a major concern for users, and unless the value of an application is extremely high (e.g., emergency response), many choose to opt-out. Fortunately, there are a number of methods for improving the privacy and anonymity of users, and we discuss them in detail in the coming chapter.

· · · ·

CHAPTER 9

# Challenges and Opportunities

This lecture has presented an overview of the most common technologies and techniques that are used to estimate location. Figure 9.1 summarizes the accuracy and coverage of these technologies. We close this lecture with a discussion of some of the issues which still present challenges to the developers of location-enhanced applications and systems. A number of these issues, such as privacy and the need for unified location abstractions, represent fertile research areas and have opportunities for systems, networking, and human–computer interaction innovation. We also present our perspective on coming commercial location technologies and how they will improve accuracy, coverage, and power consumption.

## 9.1 PRIVACY

*"Orwellian Dream Come True: A Badge That Pinpoints You."*

That was the title of the 1992 New York Times article that first described the Active Badge system in the popular press. This charged statement reflects the reality that, for many people, the privacy concerns about location technology far outweigh the perceived value of the applications it enables. This statement, of course, puts all location technologies in the same privacy bucket, and as we already learned, location technologies offer a variety of different degrees of privacy preservation. But nevertheless, it illustrates that location privacy is a hot-button, emotional issue that has to be taken into account if a system or application is to succeed.

One fundamental issue is who controls the system doing the location estimation. Client-side localization, where the user owns or controls the listener and the beacons are in the infrastructure, is far preferable to the reverse. Fortunately, there are client-side variants of nearly all of the technologies we have described. In the event that infrastructure does control or compute the location estimates, application developers need to consider the roles of the users in the deployment environment and their expectations for privacy. A hospital patient, for example, has little expectation of location privacy and may take comfort in the knowledge that the nurses' station can locate him at a moment's

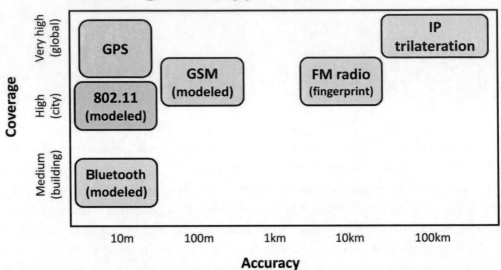

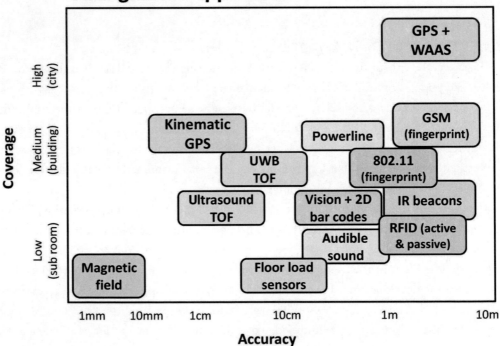

**FIGURE 9.1:** Accuracy and coverage of the technologies we have covered in this lecture.

notice. Employees, on the other hand, do not derive the same sense of security knowing that their boss can track their every movement at work.

Independent from trust of the location system itself, some application domains, such as social networking, emergency response, and location-based search involve voluntary disclosure of location. This presents another challenge for developers: ensuring that an application includes disclosure options that support the user's desired practice. Emergency response is at one end of the spectrum, and the user is nearly always motivated to disclose location information as accurately and quickly as possible. For location-based search, one can imagine a user not trusting a search provider and wanting to give blurred or false information. In practice, this has not emerged as a location-based search requirement, perhaps because of the legal recourse users have if a commercial content provider were to misuse their information. Social networking applications are more nuanced. For example, Consolvo et al. found in a study designed to determine the willingness of users to disclose their current location to people they know (e.g., spouse, coworkers, manager, friends) that the information disclosed depended on the relationship of the user with the other person, the user's current activity, and both the user and the other person's current location [29]. For instance, users were more likely to disclose exact addresses, as opposed to more vague location information such as the name of the city, to acquaintances that lived in the same city than to those that lived out-of-state. These users felt that the more detailed location information provided little extra benefit to the out-of-state acquaintance and was therefore not worth disclosing. Iachello et al. identified, in their study of a social disclosure application involving teenagers and parents, the use of four different deception techniques used for inaccurate disclosure: responses can be *delayed* and not answered immediately; responses can be *time shifted* (representing the user's location in the past or future); requests are often *ignored*; finally, responses can be *explicitly deceptive* and disclose an entirely inaccurate location [75]. Their conclusion was that it was essential for any social location disclosure application to support plausible deniability and the ability to "stretch the truth." The larger design lesson is that if a system or application does not support the preexisting expectation or social practice of the users, it will quickly fall into disuse.

*Anonymity* is often suggested as a way to improve the privacy of a location system. Weak anonymity can be offered in the form of a unique identifier (such as a badge ID or a MAC address) that is used in lieu of a person's name. A stronger form of anonymity is to give users pseudonyms that are used for only a single day or session. In other situations, the technology itself yields true anonymity such as motion sensors or smart floors that infer that someone is present, but not who. In general, anonymity is a poor substitute for real privacy, and this is especially true for location information. Unique IDs are easily reassociated with an individual's identity, as location traces routinely lead back to homes and offices. Pang et al. have shown that temporary pseudonyms offer little anonymity because of the other information leaked over a wireless network [125]. Assuming true anonymity can be achieved, there is still a threat: a mugger, for example, may not know who is in the park late at night, but they know *someone* is. Gruteser and Grunwald address this threat with a technique to allow a group of users to combine their location estimates together for increased anonymity. Their algorithm allows mobile users to achieve what they call *k-anonymity* in a location system by blurring their location estimate until it includes the location of at least *k* users [60]. The result is that in a crowded plaza, you may disclose your identity with meter resolution, while on a late-night walk, your estimates may automatically blur to neighborhood granularity.

## 9.2   "PLACE" TECHNOLOGIES

The most widely deployed location systems produce coordinate-based location estimates such as latitude and longitude. These estimates are an excellent match to navigation and emergency response applications that require either absolute locations or the ability to compute the geometric relationship between locations. Other classes of applications, such as social networking, are less well served by this type of location estimate. When did you last hear someone say something of the form: "Gosh, you'll never guess who I bumped into at 43.394 north, 79.233 west!"? For these applications, place names such as "home," "work," or "the mall" carry more semantic meaning and are more valuable as location estimates than a set of coordinates.

Room-level location systems such as infrared and ultrasound beacons can support this type of location estimate by simply augmenting their broadcasts with a place name. It requires more work for wide-area systems like GPS and modeled GSM/802.11. Ashbrook and Starner developed a technique for learning and recognizing the indoor places a GPS user goes by clustering the locations where GPS coverage is lost [6]. Similarly, Hightower et al. have developed an algorithm for using stable periods of observed 802.11 and GSM signals to learn a user's important places [70]. These techniques solve part of the problem in that they can identify a set of discrete places relevant to the user from streams of continuous location estimates. Automatically naming those places has yet to be solved, and existing systems that have made use of learned places have relied on manual user labeling [75].

Automatic place naming is a high value, but likely difficult research challenge. Service such as Microsoft's Virtual Earth and Google Maps contain information about the commercial services at or near a given set of coordinates. Cooperation schemes like those for collaborative image labeling could allow any manual naming or corrections to be shared with others. Ultimately, the problem remains difficult though, because of the personal and multi-dimensional nature of a place. One location can be, for example, *pizza parlor*, *commercial building*, *work*, *restaurant*, *rental property*, and *teen hangout* all at the same time. Questions about how to represent and use place names and how to build and use name ontologies still remain to be answered.

## 9.3 SYSTEM SUPPORT FOR LOCATION

As Figure 9.1 illustrates, there is no single location technology that is good for every situation and provides high accuracy and universal coverage. The implication is that multiple location systems are likely to coexist with applications relying on combinations of technologies based on their usage context (e.g., GPS while outdoors and WiFi fingerprinting while indoors). The lack of a homogeneous location service, however, poses a significant challenge to the development of location-aware applications which, at present, require technology-specific code for each location-sensing system in use by the application. A technology-independent application programming interface (API) is needed to shield location-aware application programmers from the nuisance of extracting location information from low-level sensors. Such an API would let applications take advantage of different location-sensing technologies by transparently fusing inputs from all available location sources.

The multiuse nature of our mobile devices raises the need for a service or library to help users manage their location information. Emergency response disclosures, for example, likely should not be blurred, anonymized, or delayed, nor should they require user confirmation. Location requests from local advertisers might be completely ignored, while social disclosure might be configured for case-by-case confirmation. The Confab architecture is an example of a toolkit that allows end-user management of location information in applications [73]. Thus far, no commercial operating systems or devices include a "location management" feature that allows these sorts of preferences to be configured.

Location awareness can also play an important role in the management of system resources. Many system functions, such as wireless communication, power management, fast wakeup, security, and authentication could be improved by using location information. Unfortunately, little has been realized or even demonstrated in practice. This is likely not due to a shortage in perceived value, but because of the lack of a high-coverage, low-power location technology to serve as the input to location-enhanced system services. The availability of near-ubiquitous low-power location via GSM/802.11/GPS fusion should eliminate this barrier and should set up location-enhanced system management as a fertile research area in the coming years.

## 9.4 TECHNOLOGIES ON THE HORIZON

A number of innovations are in commercialization that will improve the accuracy and availability of location estimation. We briefly discuss our predictions of likely changes in localization technology over the next 5 to 10 years.

Between now and 2014, the next generation of GPS satellites are being deployed. These include a second civilian code that will allow commercial GPS units to eliminate error due to ionosphere delay. This alone will likely halve the median GPS error from approximately 10 m down to

5. Once Europe's Galileo comes online, combined Galileo/GPS units should be able to provide median accuracies of 1.5 m. If time-of-flight location estimation using digital TV signals becomes popular, expect combined GPS/TV units to be available for the indoor coverage they add to GPS. In the coming years, single-chip GPS implementations will become the norm, and they should shrink in both size and power consumption. This should enable all but the most impoverished and inexpensive computational devices to include GPS, if their use merits it.

Location estimation with wireless networking signals has a less clear evolution path than GPS. While major players like Google and Microsoft have tested them in their flagship mapping products, client-side 802.11 and GSM/CDMA location estimation are not yet mainstream. The largest hurdle for 802.11 location is the emergence of a technology/business model pair that will yield a service to provide users with an up-to-date access point map. The 802.11n specification attempts to provide some standard-driven relief to this problem by including a location field in 802.11 access point advertisements. If this field was widely used, the need for an AP map goes away, as each observed AP would disclose its location. However, given the prevalence of deployed access points with default SSIDs of "netgear" and "linksys," we are skeptical that this AP location field will be widely used. The future of client-side GSM/CDMA location is dependent on the emergence of handsets that offer cell tower scan information to user-level programs. Carriers currently have little incentive to make this available, but more programmatically open phones such as those based on Microsoft's Windows Mobile or the open Google phone specification may pressure carriers to open phone capabilities in general.

Like GPS implementations, wireless networking chips will continue to shrink, making small, worn devices that can locate themselves and communicate increasingly feasible. A good example is the relatively inexpensive EyeFi SD card shown in Figure 9.2. This device fits 2 GB of storage, an 802.11b radio and the logic to upload photos into a SD form factor. If this is what we can buy today, tiny wirelessly enabled objects like smart jewelry and eventually smart dust seem likely.

Worldwide interoperability for microwave access (WiMax) and ultra-wide band (UWB) are two emerging wireless networking technologies that occupy different parts of the coverage/accuracy spectrum than 802.11 or GSM/CDMA. WiMax is intended as a last-mile solution to help broadband service providers connect customers to the network. Implementations of WiMax clients for mobile devices are now available, and with a range of 1–2 km, WiMax may provide more accurate location estimation than GSM/CDMA in rural and suburban areas. At the other end of the spectrum, the Wireless USB standard is an effort to use ultra-wide band radio as a wire replacement. Wireless USB promises bandwidth of around 500 MB/s with a 3-m range and very low power consumption. Used as a location technology, wireless USB signals should provide extremely accurate location estimates and, if the standard takes root, could be a common source of subroom location estimates for homes and offices.

**FIGURE 9.2:** The EyeFi SD card includes 2 GB of storage and an 802.11 radio and the logic to automatically upload photos to the Internet or a PC.

## 9.5   CONCLUSIONS

The technology trend that has thus far gone unsaid is that mobile devices will continue to shrink in size and price while growing in capability and usefulness to people. This will only increase the breadth and depth of interaction that users have with location-aware devices and services in the future. In this lecture, we have provided a technical introduction to the many techniques that have been developed for location estimation. While there is no single technology that provides inexpensive, accurate, high-coverage location, the improvements in GPS, the increasing density and variety of wireless networking, and the new location fusion algorithms hold the promise of a hybrid location system with just these qualities.

•   •   •   •

# References

[1] 911 Services. www.fcc.gov/pshs/911.

[2] Abowd, G., Atkeson, C., Hong, J., Long, S., Kooper, R., and Pinkerton, M., "Cyberguide: A mobile context-aware tour guide," *Journal Wireless Networks*, 3(5), October 1997, pp. 421–433.

[3] Abowd, G., Battestini, A., and O'Connell, T., "The location service: a framework for handling multiple location sensing technologies," *GVU Technical Report;GIT-GVU-04-20*, 2002.

[4] ASCC and Versus Information Systems ascc-inc.com/HealthCare/Versus.htm.

[5] Ascension Technology Corporation, www.ascension-tech.com.

[6] Ashbrook, D., and Starner, T., "Using GPS to learn significant locations and predict movement across multiple users," *Personal and Ubiquitous Computing*, 7, 2003, pp. 275–286. doi:10.1007/s00779-003-0240-0

[7] Axelrad, P., and R.G. Brown., GPS navigation algorithms. Vol. I, Chap. 9 in Global positioning system: theory and applications, edited by Bradford W. Parkinson and James J. Spilker. Cambridge, Massachusetts: American Institute of Aeronautics and Astronautics, Inc., 1996.

[8] Bahl, P., and Padmanabhan, V., "RADAR: An in-building RF-based user location and tracking system," in *Proceedings of IEEE INFOCOM*, Tel-Aviv, Israel, 2000, pp. 775–784. doi:10.1109/INFCOM.2000.832252

[9] Barkhuus, L., Chalmers, M., Tennent, P., Hall, M., Bell, M., Sherwood, S., and Brown, B., "Picking pockets on the lawn: The development of tactics and strategies in a mobile game," *Proceedings UbiComp 2005: Ubiquitous Computing: 7th International Conference*, Tokyo, Japan, 2005, pp. 358–374.

[10] Battiti, R., Nhat, T.L., and Villani, A., "Location-aware computing: A neural network model for determining location in wireless lans," Technical Report DIT-5, Universita di Trento, Dipartimento di Informatica e Telecomunicazioni, 2002.

[11] Beckman, C., Consolvo, S., and LaMarca, A., "Some assembly required: Supporting end-user sensor installation in domestic computing environments," *Proceedings of the 6th International Conference on Ubiquitous Computing*, 2004, pp. 102–124.

[12] Benford, S., Anastasi, R., Flintham, M., Greenhalgh, C., Tandavanitj, N., Adams, M., and Row-Farr, J., "Coping with uncertainty in a location-based game," *IEEE Pervasive Computing*, 2(3), 2003, pp. 34–41. doi:10.1109/MPRV.2003.1228525

[13] Bhasker, E.S., Brown, S.W., and Griswold, W.G., "Employing user feedback for fast, accurate, low-maintenance geolocationing," *Proceedings of the Second IEEE International Conference on Pervasive Computing and Communications (PerCom 2004), March 2004*, Orlando, FL, 2004, pp. 111–120. doi:10.1109/PERCOM.2004.1276850

[14] Bian, X., Abowd, G.D., and Rehg, J.M., "Using sound source localization in a home environment," *Proceedings of The 3rd International Conference on Pervasive Computing*, Munich, Germany, 2005, pp. 19–36.

[15] Borkowski, J., Niemela, J., and Lempiainen, J., "Performance of cell ID+RTT hybrid positioning method for UMTS radio networks," *Fifth European Wireless Conference*, 2004. doi:10.1109/VETECF.2004.1404724

[16] Borriello, G., Liu, A., Offer, T., Palistrant, C., and Sharp, R., "WALRUS: wireless acoustic location with room-level resolution using ultrasound," *Proceedings of the 3rd International Conference on Mobile Systems, Applications, and Services*, 2005, pp. 191–203. doi:10.1145/1067170.1067191

[17] Brumitt, B., Meyers, B., Krumm, J., Kern, A., and Shafer, S., "EasyLiving: Technologies for intelligent environments," *Handheld and Ubiquitous Computing, Second International Symposium*, 2000, pp. 12–29. doi:10.1007/3-540-39959-3_2

[18] Burrell, J., Gay, G., Kubo, K., and Farina, N., "Context-aware computing: A test case," *4th International Conference on Ubiquitous Computing*, 2002, pp. 1–15. doi:10.1007/3-540-45809-3_1

[19] Caffery, J.J., and Stuber, G.L., "Overview of radiolocation in CDMA cellular systems," *IEEE Communications Magazine*, 36(4), 1998, pp. 38–45. doi:10.1109/35.667411

[20] Carter, S., Churchill, E., Denoue, L., Helfman, J., and Nelson, L., "Digital graffiti: Public annotation of multimedia content," *CHI '04 Extended Abstracts on Human Factors in Computing Systems*, 2004, pp. 1207–1210.

[21] Castro, P., Chiu, P., Kremenek, T., and Muntz, R., "A probabilistic location service for wireless network environments," *Proceedings of Ubiquitous Computing, Third International Conference*, Atlanta, GA, September 2001, pp. 18–34. doi:10.1007/3-540-45427-6_3

[22] Chandrasekaran, G., Ergin, M.A., Gruteser, M., and Martin, R.P., "Bootstrapping a location service through geocoded postal addresses," *Proceedings of 3rd International Symposium on Location and Context-Awareness (LoCA)*, 2007. doi:10.1007/978-3-540-75160-1_1

[23] Chen, M.Y., Sohn, T., Chmelev, D., Haehnel, D., Hightower, J., Hughes, J., LaMarca, A., Potter, F., Smith, I., and Varshavsky, A., "Practical metropolitan-scale positioning for GSM

phones," *Eighth International Conference on Ubiquitous Computing*, Irvine, California, 2006, pp. 225–242.

[24] Cheng, Y.C., Chawathe, Y., LaMarca, A., and Krumm, J., "Accuracy characterization for metropolitan-scale Wi-Fi localization," *Third International Conference on Mobile Systems, Applications, and Services (MobiSys 2005)*, 2005, pp. 233–245. doi:10.1145/1067170 .1067195

[25] Cheok, D., Goh, K., Farbiz, W., Fong., S., Li, S., and Yang, Z., "Human Pacman: A mobile, wide-area entertainment system based on physical, social, and ubiquitous computing," *Personal and Ubiquitous Computing*, 8(2), 2001, pp. 71–81. doi:10.1007/s00779-004-0267-x

[26] Cheverst, K., Davies, N., Mitchell, K., Friday, A., and Efstratiou, C., "Developing a context-aware electronic tourist guide: Some issues and experiences," *Proceedings of the SIGCHI Conference on Human Factors in Computing Systems*, April 2000, pp. 17–24. doi:10.1145/ 332040.332047

[27] Collins, R., Lipton, A., Fujiyoshi, H., and Kanade, T., "Algorithms for cooperative multisensor surveillance," *Proceedings of the IEEE*, 89(10), 2001, pp. 1456–1477. doi:10.1109/5.959341

[28] Comaniciu, D., Ramesh, V., and Meer, P., "Kernel-based object tracking," *IEEE Transactions on Pattern Analysis and Machine Intelligence*, 25(5), 2003, pp. 564–575. doi:10.1109/ TPAMI.2003.1195991

[29] Consolvo, S., Smith, I., Matthews, T., LaMarca, A., Tabert, J., and Powledge, P.,"Location disclosure to social relations: Why, when, & what people want to share," *Proceedings of CHI '05*, April, 2005. doi:10.1145/1054972.1054985

[30] DARPA Grand Challenge, www.darpa.mil/grandchallenge.

[31] Data and Products. http://igscb.jpl.nasa.gov/components/prods_cb.html. Prod. The International GNSS Service. November 7, 2007.

[32] Davidson, A., "Real-time simultaneous localisation and mapping with a single camera," *in Proc. IEEE Int. Conf. Computer Vision, ICCV-2003*, 2003, pp. 1403–1410.

[33] de Ipina, D.L., Mendonça, P.R.S., and Hopper, A., "TRIP: A low-cost vision-based location system for ubiquitous computing," *Personal and Ubiquitous Computing*, 6(3), 2002, pp. 206–219.

[34] Djuknic, G.M., and Richton, R.E., "Geolocation and assisted GPS," *IEEE Computer*, 34(2), 2001, 123–125. doi:10.1109/2.901174

[35] Dodgeball.com: Social Mobile Software. www.dodgeball.com

[36] Doucet, A., and De Freitas, N., *Sequential Monte Carlo in practice*, New York, Springer-Verlag, 2001.

[37] Doucet, A., Godsill, S., and Andrieu, C., "On sequential Monte Carlo sampling methods for Bayesian filtering," *Statistics and Computing*, 10(3), 2000, pp. 197–208.

[38]   Drane, C., Macnaughtan, M., and Scott, C., "Positioning GSM Telephones," *IEEE Communications Magazine*, 36(4), 46–59. doi:10.1109/35.667413

[39]   Eagle, N., and Pentland, A., "Reality mining: Sensing complex social systems," *Personal and Ubiquitous Computing*, 10(4), 2006, pp. 255–268.

[40]   Ekahau, Inc, http://www.ekahau.com.

[41]   El-Rabbany, Ahmed. Introduction to the global positioning system. Norwood, Artech House Publishers, 2006. doi:10.1007/s00779-005-0046-3

[42]   Enge, P., and Misra P., "Scanning the issue/technology," *Proceedings of the IEEE*, 87(1), January 1999.

[43]   Feng, Y.M., "Future GNSS performance," *GPS World*, May 2005.

[44]   Ferris, B., Haehnel, D., and Fox, D., "Gaussian processes for signal strength-based location estimation," *Proceedings of Robotics: Science and Systems*, Philadelphia, PA, August 16–19, 2006.

[45]   Fox, D., "Adapting the sample size in particle filters through KLD-Sampling," *International Journal of Robotics Research*, 22, 2003. doi:10.1177/0278364903022012001

[46]   Fox, D., Burgard, W., and Thrun, S., "Active Markov localization for mobile robots," *Robotics and Autonomous Systems*, 25(3), 1998, pp. 195–207. doi:10.1016/S0921-8890(98)00049-9

[47]   Fox, D., Hightower, J., Liao, L., Schulz, D., and Borriello, G., "Bayesian filters for location estimation," *IEEE Pervasive Computing*, 2, September 2003, pp. 24–33.

[48]   Frequently Asked Questions. http://pnt.gov/public/faq.shtml#accurate. (accessed November 15, 2007).

[49]   Fritz, G., Seifert, C., and Paletta, L., "A mobile vision system for urban detection with informative local descriptors," *ICVS '06: Proceedings of the Fourth IEEE Intl Conference on Computer Vision Systems*, 2006. doi:10.1109/ICVS.2006.5

[50]   Fukumoto, M., and Shinagawa, M., "CarpetLAN: A novel indoor wireless(-like) networking and positioning system," *Ubiquitous Computing, 7th International Conference*, 2005, pp. 1–18.

[51]   Garg, S., Kappes, M., and Mani, M., "Wireless access server for quality of service and location based access control in 802.11 networks," *Proceedings of Computers and Communications*, 2002, pp. 819–824. doi:10.1109/ISCC.2002.1021767

[52]   Getting, I.A., "The global positioning system," IEEE Spectrum 30(12), December 1993, pp. 36–47. doi:10.1109/6.272176

[53]   GLONASS Constellation Status. http://www.glonass-ianc.rsa.ru/pls/htmldb/f?p=202:20:12676890526720372432::NO (accessed November 16, 2007).

[54]   Gmaps Pedometer. www.gmap-pedometer.com

[55]   Gordon, N., Salmond, D., and Smith, A., "Novel approach to nonlinear and non-Gaussian Bayesian state estimation," *IEEE Proc. F*, 140, 1993, pp. 107–113.

[56] GPS Modernization Fact Sheet, The National Space-Based Positioning, Navigation, and Timing (PNT) Executive Committee. November 16, 2007. http://pnt.gov/public/docs/2006-01-modernization.pdf.

[57] GPS-Enabled Location-Based Services (LBS) Subscribers Will Total 315 Million in Five Years, (2006, September 26). http://www.abiresearch.com/abiprdisplay.jsp?pressid=731

[58] Griswold, W.G., Shanahan, P., Brown, S.W., Boyer, R. Ratto, M., Shapiro, R.B., and Truong, T.M., "ActiveCampus: Experiments in community-oriented ubiquitous computing," *IEEE Computer*, 37(10), October 2004, pp. 73–81.

[59] Griswold, W.G., Shanahan, P., Brown, S.W., Boyer, R., Ratto, M., Shapiro, R.B., and Truong, T.M., "ActiveCampus: Experiments in community-oriented ubiquitous computing," *IEEE Computer*, 37(10), October 2004, pp. 73–81.

[60] Gruteser, M., and Grunwald, D., "Anonymous usage of location-based services through spatial and temporal cloaking," *Proceedings of the 1st international Conference on Mobile Systems, Applications and Services (MobiSys '03)*, 2003, pp. 31–42.

[61] Gustafsson, F., and Gunnarsson, F., "Mobile positioning using wireless networks," *IEEE Signal Processing Magazine*, 2005, pp. 41–53.

[62] Gustafsson, F. Gunnarsson, F., Bergman, N., Forssell, U., Jansson, J., Karlsson, R., and Nordlund, P.J., "Particle filters for positioning, navigation, and tracking," *IEEE Transactions on Signal Processing*, 50(2), February 2002, pp. 425–437.

[63] Haeberlen, A., Flannery, E., Ladd, A.M., Rudys, A., Wallach, D.S., and Kavraki, L.E., "Practical robust localization over large-scale 802.11 wireless networks," *Proceedings of ACM MobiCom '04*, Philadelphia, PA, September 2004, pp. 70–84. doi:10.1145/1023720.1023728

[64] Hähnel, D, Burgard, W, Fox, D., Fishkin K, and Philipose, M., "Mapping and localization with RFID technology," *2004 IEEE International Conference on Robotics and Automation*, 2004, pp. 1015–1020. doi:10.1109/ROBOT.2004.1307283

[65] Haritaoglu, I., Harwood, D., and Davis, L.S., "W4: Real-time surveillance of people and their activities," *Transactions on Pattern Analysis and Machine Intelligence*, 22(8), August 2000, pp. 809–830. doi:10.1109/34.868683

[66] Harter, A., Hopper, A., Steggles, P., Ward, A., and Webster, P., "The anatomy of a context-aware application," *Wireless Networks*, 8(2–3), 2002, pp. 187–197.

[67] Hightower, J., and Borriello, G., "Particle filters for location estimation in ubiquitous computing: A case study," *Proceedings UbiComp 2004: Ubiquitous Computing: 6th International Conference*, Nottingham, UK, September, *2004*.

[68] Hightower, J., Borriello, G., "Location systems for ubiquitous computing," *Computer*, 34(8), August 2001, pp. 57–66. doi:10.1109/2.940014

[69] Hightower, J., Borriello, G., and Want, R., "SpotON: An indoor 3D location sensing technology based on RF signal strength," *Tech. Rep. #2000-02-02*, University of Washington, February 2000.

[70] Hightower, J., Consolvo, S., LaMarca, A., Smith, I., and Hughes, J., "Learning and recognizing the places we go," *Proceedings of UbiComp '05*, 2005, pp. 159–176.

[71] Hightower, J., *The location stack*, : Ph.D. Thesis, University of Washington, 2004.

[72] Hile, H., and Borriello, G., "Information overlay for camera phones in indoor environments," *Proceedings of 3rd International Symposium on Location and Context-Awareness (LoCA)*, 2007. doi:10.1007/978-3-540-75160-1_5

[73] Hong, J.I., and Landay, J.A, "An architecture for privacy-sensitive ubiquitous computing," *Proceedings of the 2nd international Conference on Mobile Systems, Applications, and Services (MobiSys '04)*, 2004, pp. 177–189. doi:10.1145/990064.990087

[74] Huwedi, A., Steinhaus, P., and Dillmann, R., "Autonomous feature-based exploration using multi-sensors," *2006 IEEE International Conference on Multisensor Fusion and Integration for Intelligent Systems*, September 2006, pp. 456–461.

[75] Iachello, G., Smith, I., Consolvo, S., Abowd, G., Hughes, J., Howard, J., Potter, F.,Scott, J., Sohn, T., Hightower, J., and LaMarca, A., "Control, deception, and communication:Evaluating the deployment of a location-enhanced messaging service," *Proceedings of Ubicomp '05*, 2005, pp. 213–231.

[76] Internet-based Global Differential GPS. http://gipsy.jpl.nasa.gov/igdg/system/. Prod. Jet Propulsion Laboratory. November 7, 2007.

[77] JSR-179: The Java Location API. www.jcp.org/en/jsr/detail?id=179

[78] Kaplan, E.D., "Introduction," Chap. 1 in Understanding GPS principles and applications, edited by Elliot D. Kaplan and Christopher J. Hegarty, Artech House, 2005, pp. 1–20.

[79] Karalar, T.C., and Rabaey, J., "An RF ToF based ranging implementation for sensor networks," *IEEE International Conference on Communications*, 2006, pp. 3347–3352.

[80] Katz-Bassett, E., John, J.P., Krishnamurthy, A., Wetherall, D., Anderson, T., and Chawathe, Y., "Towards IP geolocation using delay and topology measurements," *Proceedings of the 6th ACM SIGCOMM Conference on internet Measurement*, 2006, pp. 71–84. doi:10.1145/1177080.1177090

[81] Kim, D., and Langley, R.B., "GPS ambiguity resolution and validation: Methodologies, trends and issues," *7th GNSS Workshop—International Symposium on GPS/GNSS*, Seoul, Korea, 2000.

[82] Kim, M., Fielding, J.J., and Kotz, D., "Risks of using AP locations discovered through war driving," *Proceedings of Pervasive Computing, 4th International Conference, PERVASIVE 2006*, Dublin, Ireland, May 7–10, 2006, pp. 67–82. doi:10.1007/11748625_5

[83] King, T., Haenselmann, T., and Effelsberg, W., "Deployment, calibration, and measurement factors for position errors in 802.11-based indoor positioning systems," *Proceedings of 3rd International Symposium on Location and Context-Awareness (LoCA)*, 2007. doi:10.1007/978-3-540-75160-1_2

[84] Kismet, http://www.kismetwireless.net/.

[85] Kleusberg, A., and Langley, R.B., "The limitations of GPS," *GPS World*, 1(2), March/April 1990, pp. 50–52.

[86] Köhler, M., Patel, S., Summet, J., Stuntebeck, E., and Abowd, G., "TrackSense: Infrastructure free precise indoor positioning using projected patterns," *Pervasive Computing, 5th International Conference*, 2007, pp. 334–350. doi:10.1007/978-3-540-72037-9_20

[87] Kosaka, A., and Pan, J., "Purdue experiments in model-based vision for hallway navigation," *IEEE/RSJ International Conference on Intelligent Robots and Systems (IROS)*, 1995, pp. 87–96.

[88] Krumm, J., and Hinkley, K, "The NearMe wireless proximity server," *Proceedings UbiComp 2004: Ubiquitous Computing: 6th International Conference*, Nottingham, UK, September 2004, pp. 283–300.

[89] Krumm, J., and Horvitz, E., "LOCADIO: Inferring motion and location from Wi-Fi signal strengths," *1st Annual International Conference on Mobile and Ubiquitous Systems (MobiQuitous 2004)*, Cambridge, MA, 2004, pp. 4–13. doi:10.1109/MOBIQ.2004.1331705

[90] Krumm, J., and Horvitz, E., "Predestination: Inferring destinations from partial trajectories," *UbiComp 2006: Ubiquitous Computing, 8th International Conference*, Orange County, CA, September 17–21, 2006, pp. 243–260. doi:10.1007/11853565_15

[91] Krumm, J., Cermak, G., and Horvitz, E., "*RightSPOT: A novel sense of location for a smart personal object*," *Ubiquitous Computing, 5th International Conference*, 2003, pp. 36–43.

[92] Laasonen, K., Raento, M., and Toivonen, H., "Adaptive on-device location recognition," *2nd International Conference on Pervasive Computing*. Vienna, Austria, 2004, pp. 287–304.

[93] Ladd, A.M., Bekris, K.E., Rudys, A., Marccau, G., and Kavraki, L.E., "Robotics-based location sensing using wireless ethernet," *Proceedings of ACM MobiCom*, September 2002, pp. 227–238. doi:10.1145/570645.570674

[94] Laitinen, H., Lahteenmaki, J., and Nordstrom, T., "Database Correlation Method for GSM Location," *53rd IEEE Vehicular Technology Conference*, Rhodes, Greece, 2001, pp. 2501–2508. doi:10.1109/VETECS.2001.944052

[95] LaMarca, A., Chawathe, Y., Consolvo, S., Hightower, J., Smith, I., Scott, J., Sohn, T., Howard, J., Hughes, J., Potter, F., Tabert, J., Powledge, P., Borriello, G., and Schilit, B., "Place Lab: Device positioning using radio beacons in the wild," *Proceedings of Pervasive*

*Computing, Third International Conference, PERVASIVE 2005*, Munich, Germany, May 8–13, 2005, pp. 116–133.

[96]    LaMarca, A., Hightower, J., Smith, I., and Consolvo, S., "Self-Mapping in 802.11 Location Systems," *Proceedings of 7th International Conference, UbiComp 2005*, Tokyo, Japan, September 11–14, 2005, pp. 87–104.

[97]    Langley, R.B., "Smaller and smaller: The evolution of the GPS receiver," *GPS World Maganize*, 11(4), April 2000, pp. 54–58.

[98]    Langley, Richard B., "Why is the GPS Signal So Complex," *GPS World Magazine*, 1(3), May/June 1990, pp. 56–59.

[99]    Letchner, J., Fox, D., and LaMarca, A., "Large-scale localization from wireless signal strength," *Proceedings of the National Conference on Artificial Intelligence (AAAI)*, 2005, pp. 15–20.

[100]   Levy, L.J., "The Kalman filter: Navigation's integration workhorse," *GPS World*, 8(9), 1996, pp. 65–71.

[101]   Li, X., and Bar-Shalom, Y., "Multiple-model estimation with variable structure," *IEEE Transactions on Automatic Control*, 41(4), April 1996, pp. 478–493.

[102]   Li, Y., Hong, J.I., and Landay, J.A., "Topiary: A tool for prototyping location-enhanced applications," *Proceedings of the 17th Annual ACM Symposium on User interface Software and Technology*, October 2004, pp. 217–226. doi:10.1145/1029632.1029671

[103]   Liu, X., Corner, M., and Shenoy, P., "Ferret: RFID localization for pervasive multimedia," *8th International Conference, UbiComp 2006*, September 2006, pp. 422–440. doi:10.1007/11853565_25

[104]   Lowe, D., "Object recognition from local scale-invariant features," *Proc. 7th Int. Conf. Computer Vision*, 1999, pp. 1150–1157.

[105]   Madhavapeddy, A., and Tse, A., "A study of bluetooth propagation using accurate indoor location mapping," *UbiComp 2005: Ubiquitous Computing*, 2005, pp. 105–122.

[106]   Madhavapeddy, A., Scott, D., and Sharp, R., "Context-aware computing with sound," *Ubiquitous Computing, 5th International Conference*, Seattle, WA, October 2003, pp. 315–332.

[107]   Mansley, K., Beresford, A., and Scott, D., "The carrot approach: encouraging use of location systems," *UbiComp 2004: Ubiquitous Computing*, 2004, pp. 366–383.

[108]   Marmasse, N., and Schmandt, C., "A user-centered location model" *Personal and Ubiquitous Computing*, 6, 2002, pp. 318–321.

[109]   Maybeck, P.S., *Stochastic Models, Estimation and Control*, I, Academic Press, New York, 1979.

[110]   McDonald, K.D., and Hegarty, C., "Post-modernization GPS performance capabilities," *Proceedings of the 56th Institute on Navigation (ION) Annual Meeting*. San Diego, CA, 2000, pp. 242–249.

[111] McGuire, M., Plataniotis, K.N., and Venestsanopoulos, A.N., "Location of mobile terminals using time measurements and survey points," *IEEE Transaction on Vehicular Technology*, 52(4), 2003, pp. 999–1011. doi:10.1109/TVT.2003.814222

[112] Misra, P., Burke B.P., and Pratt, M.M., "GPS performance in navigation," *Proceedings of the IEEE*, 87(1), January 1999, pp. 65–85.

[113] Misra, P., and and Enge, P., *The Global Positioning System: Signals, Measurements and Performance, 2nd ed.*, Ganga-Jamuna Press, 2006.

[114] Montemerlo, M., Thrun, S., Koller, D., Wegbreit, B., "FastSLAM: A factored solution to the simultaneous localization and mapping problem," *Eighteenth national conference on Artificial intelligence*, 2002, pp. 593–598.

[115] Muller, H.L., McCarthy, M.R., and Randell, C., "Particle filters for position sensing with asynchronous ultrasonic beacons," *Location- and Context-Awareness, Second International Workshop, (LoCA 2006)*, Dublin, Ireland, May 2006, pp. 1–13. doi:10.1109/5.736342

[116] Navizon: Peer-to-peer wireless positioning, www.navizon.com.

[117] NetGeo—Internet Geography Intelligence. www.netgeo.com.

[118] Netstumbler.com, http://www.netstumbler.com.

[119] Ni, L.M., Liu, Y., Lau, Y.C., and Patil, A., "LANDMARC: Indoor location sensing using active RFID," *Journal Wireless Networks*, 10(6), 2004, pp. 1572–8196. doi:10.1017/S0373463398007747

[120] OnStar. www.onstar.com.

[121] Orr, J., and Abowd, G., "The smart floor: A mechanism for natural user identification and tracking," *Conference on Human Factors in Computing Systems (CHI 2000)*, April, 2000.

[122] Otsason, V., Varshavsky, A., LaMarca, A., and de Lara, E., "Accurate GSM indoor localization," *Seventh International Conference on Ubiquitous Computing*, Tokyo, Japan, 2005.

[123] Pace, S., Frost, G., Lachow, I., Frelinger, D., Fossum, D., Wassem, D.K., and Pinto, M., "GPS history, chronology, and budgets," *Chap. Appendix B in The Global Positioning System, Assessing National Policies*, RAND Corporation, 1994, pp. 237–270.

[124] Padamanabhan, V.N., and Lakshminarayanan S., "Determining the geographic location of Internet hosts," *SIGMETRICS Perform. Eval. Rev.*, 29(1), June 2001, pp. 324–325. doi:10.1145/384268.378814

[125] Pang, J., Greenstein, B., Gummadi, R., Seshan, S., and Wetherall, D., "802.11 user fingerprinting," *Proceedings of ACM Mobicom 2007*, 2007.

[126] Paradiso, J., Abler, C., Hsiao, K., and Reynolds, M., "The magic carpet: Physical sensing for immersive environments," *Extended Abstracts of CHI'97*, Atlanta, GA, March 1997, pp. 277–278.

[127]   Parkinson, B.W., "Introcution and heritage of NAVSTAR, the global positioning system, Vol. I," in *Global Positioning System: Theory and Applications*, edited by Bradford W. Parkinson, James J. Spilker Jr., Penina Axelrad and Per Enge, Cambridge: American Institute of Aeronautics and Astronautics, Inc., 1996, pp. 3–28.

[128]   Patel, S., Truong, K., and Abowd, G., "PowerLine positioning: A practical sub-room-level indoor location system for domestic use," *Ubiquitous Computing 8th International Conference*, 2006, pp. 441–458. doi:10.1007/11853565_26

[129]   Patterson, D.J., Liao, L., Gajos, K., Collier, M., Livic, N., Olson, K., Wang, S., Fox, D., and Kautz, H.A., "Opportunity knocks: A system to provide cognitive assistance with transportation services," *Ubiquitous Computing: 6th International Conference*, Nottingham, UK, September 2004, pp. 433–450.

[130]   Periakaruppan, R., and Nemeth, E., "GTrace—a graphical traceroute tool.," *USENIX LISA*, November 1999, pp. 69–78.

[131]   PlaceEngine. www.placeengine.com.

[132]   PlanetLab. www.planet-lab.org.

[133]   Presidential Directive: U.S. Space-Based Positioning, Navigation, and Timing Policy, The Office of Science and Technology Policy. December 15, 2004. http://www.ostp.gov/html/FactSheetSPACE-BASEDPOSITIONINGNAVIGATIONTIMING.pdf.

[134]   Privacy Observant Location System. pols.sourceforge.net.

[135]   Priyantha, N.B., Miu, A., Balakrishnan, H., and Teller, S., "The Cricket compass for context-aware mobile applications," *Proceedings of the 6th Annual International Conference on Mobile Computing and Networking*, Rome, Italy, July 2001, pp. 1–14. doi:10.1145/381677.381679

[136]   Priyantha, N.B., Chakraborty, A., and Balakrishnan, H., "The Cricket location-support system," *Proceedings of the sixth annual international conference on Mobile computing and networking*, Boston, MA, August 2000, pp. 32–43. doi:10.1145/345910.345917

[137]   Rabinowitz, M., and Spilker, J., "Positioning using the ATSC digital television signal," *Rosum Corporation Whitepaper*, Rosum, Redwood City, California, 2001.

[138]   Rekimoto, J., and Ayatsuka, Y., CyberCode: designing augmented reality environments with visual tags" *Proceedings of DARE 2000 on Designing augmented reality environments*, 2000, pp. 1–10.

[139]   Rekimoto, J., Miyaki, T., and Ishizawa, T., "LifeTag: WiFi-based continuous location logging for life pattern analysis," *Proceedings of 3rd International Symposium on Location and Context-Awareness (LoCA)*, 2007. doi:10.1007/978-3-540-75160-1_3

[140]   Resolution on the Galileo Concession Contract Negotiations, European Parliament, May 16, 2007. http://www.europarl.europa.eu/oeil/file.jsp?id=5478142.

[141] Rizos, C., Higgins, M.B., and Hewitson S., "New global navigation satellite system developments and their impact on survey service providers and surveyors," *Proceedings of SSC2005 Spatial Intelligence, Innovation and Praxis: The national biennial Conference of the Spatial Sciences Institute*, Melbourne, September 2005.

[142] Rosum Corporation, www.rosum.com.

[143] Rous, M., Lupschen, H., and Kraiss, K., "Vision-based indoor scene analysis for natural landmark detection," *IEEE International Conference on Robotics and Automation*, 2005, pp. 4642–4647. doi:10.1109/ROBOT.2005.1570836

[144] RTCM Recommendation Standards for Differential GNSS Service. Arlington, Virginia: Radio Technical Commission for Maritime Services, February 10, 2004.

[145] Russell, S.J., and Norvig, P., *Artificial Intelligence: A Modern Approach, 2nd ed.*, Prentice Hall, 2002.

[146] Saffiotti, A., "The uses of fuzzy logic in autonomous robot navigation," *Journal of Soft Computing*, 1(4), December 1997, pp. 180–197.

[147] Sakata, M., Yasumuro, Y., Imura, M., Manabe, Y., and Chihara, K., "ALTAIR: Automatic location tracking system with Active IR-tag," *Proceedings of IEEE International Conference on Multisensor Fusion and Integration for Intelligent Systems*, 2003, pp. 299–304. doi:10.1109/MFI-2003.2003.1232674

[148] Schloter, C.P., and Aghajan, H., "Wireless symbolic positioning using support vector machines," *Proceedings of the 2006 international conference on Wireless communications and mobile computing*, 2006. doi:10.1145/1143549.1143778

[149] Schmidt, A., Strohbach, M., Van Laerhoven, K., Friday, A., and Gellersen, H., "Context acquisition based on load sensing," *Proceedings of Ubicomp 2002*, September 2002.

[150] Scott, J., and Dragovic, B., "Audio location: Accurate low-cost location sensing," *Third International Conference, Pervasive 2005*, 2005, pp. 1–18.

[151] Scott, J., and Hazas, M., "User-friendly surveying techniques for location-aware systems," *Ubiquitous Computing, 5th International Conference*, Seattle, WA, October 2003, pp. 44–53.

[152] Seidel, S.Y., and Rapport, T.S., "914 Mhz path loss prediction model for indoor wireless communications in multifloored buildings," *IEEE Transactions on Antennas and Propagation*, 40, 1992, pp. 207–217. doi:10.1109/8.127405

[153] Shaw, M., "Modernization of the global positioning system," *Acta Astronautica (International Academy of Astronautics)*, 54(11), 2004, pp. 943–947. doi:10.1016/j.actaastro.2004.01.036

[154] Shyhook Wireless. skyhookwireless.com.

[155] Smailagic, A., Siewiorek, D.P., Anhalt, J., Kogan, D., and Wang, Y., "Location sensing and privacy in a context aware computing environment," *Proceedings of Pervasive Computing*, 2001.

[156]   Sohn, T., Li, K.A., Lee, G., Smith, I., Scott, J., and Griswold, W.G., "Place-Its: A study of location-based reminders on mobile phones," *Proceedings of Ubicomp 2005*, 2005, pp. 232–250.

[157]   Sohn, T., Varshavsky, A., LaMarca, A., Chen, M., Choudhury, T., Smith, I., Consolvo, S., Hightower, J., Griswold, W., and de Lara, E., "Mobility detection using everyday GSM traces," *in Proceedings of the Eighth International Conference on Ubiquitous Computing (Ubicomp 2006)*, September 2006, pp. 212–224. doi:10.1007/11853565_13

[158]   Soliman, S., Agashe, P., Fernandez, I., Vayanos, A., Gaal, P., and Oljaca, M., "gpsOne: A hybrid position location system," *6th International Symposium on Spread Spectrum Techniques and Applications*, Parsippany, New Jersey, 2000, pp. 330–335. doi:10.1109/ISSSTA.2000.878139

[159]   Song, H.L., "Automatic vehicle location in cellular communications systems," *IEEE Transactions of Vehicular Technology*, 43(04), 1994, pp. 902–908.

[160]   Statement by the Press Secretary. http://www.whitehouse.gov/news/releases/2007/09/20070918-2.html. The White House. September 18, 2007.

[161]   Steggles, P., and Gschwind, S., "The ubisense smart space," *UbiSense Corporation Whitepaper*, May, 2005.

[162]   Stenbit, J.P., "GPS SPS Performance Standard," *The National Space-Based Positioning, Navigation, and Timing (PNT) Executive Committee*, October 2001. http://pnt.gov/public/docs/SPS-2001-final.pdf.

[163]   Suwantrugul, S., Rakariyatham, P., Komolmis, T., and Sung-in, A., "A Modelling of Ionospheric Delay over Chiang Mai Province," *Proceedings of the IEEE International Symposium on Circuits and Systems (ISCAS)*, Bangkok, Thailand, May 2003.

[164]   Toyama, K., Logan, R., and Roseway, A., "Geographic location tags on digital images," in *Proceedings of the Eleventh ACM international Conference on Multimedia (MULTIMEDIA '03)*, November 2003, pp. 156–166. doi:10.1145/957013.957046

[165]   UbiSense, www.ubisense.net.

[166]   Varshavsky, A., de Lara, E., Hightower, J., LaMarca, A., and Otsason, V., "GSM indoor localization," *Pervasive and Mobile Computing Journal*, 3(6), 2007, pp. 698–720. doi:10.1016/j.pmcj.2007.07.004

[167]   Varshavsky, A., LaMarca, A., Hightower, J., and de Lara, E., "The SkyLoc floor localization system," *Fifth IEEE International Conference on Pervasive Computing and Communications*, White Plains, New York, 2007, pp. 125–134.

[168]   Vindigo, www.vindigo.com.

[169]   WAAS Performance Analysis Report. National Satellite Test Bed (NSTB), July 2007.

[170]   Want, R., "RFID explained: A primer on radio frequency identification technologies," *Synthesis Lectures on Mobile and Pervasive Computing*, 1, January 2006, pp. 1–94.

[171] Want, R., Hopper, A., Falcão, V., and Gibbons, J., "The active badge location system," *ACM Transactions on Information Systems*, January 1992, pp. 91–102. doi:10.1145/128756 .128759

[172] Ward, N., "The Status, Development and Future Role of Radio beacon Differential GNSS," *Journal of Navigation*, 51(2), 1998, pp. 152–158.

[173] Weiss, A.J., "On the accuracy of a cellular location system based on RSS measurements," *IEEE Transactions on Vehicular Technology*, 52(6), 2003, pp. 1508–1518. doi:10.1109/ TVT.2003.819613

[174] Wigle: Wireless Geographic Logging Engine, wigle.net.

[175] Youssef, A., Krumm, J., Miller, E., Cermak, G., and Horvitz, E., "Computing location from ambient FM radio signals," *2005 IEEE Wireless Communications and Networking Conference*, March 2005, pp. 824–829. doi:10.1109/WCNC.2005.1424614

[176] Youssef, M.A., and Agrawala, A., "The Horus WLAN location determination system," in *Third International Conference on Mobile Systems, Applications, and Services (MobiSys 2005)*, 2005, pp. 205–218. doi:10.1145/1067170.1067193

[177] Youssef, M.A., Agrawala, A., and Shankar, U.A., "WLAN Location determination via clustering and probability distributions," *Proceedings of the First IEEE International Conference on Pervasive Computing and Communications*, March 2003, pp. 143–150. doi:10.1109/ PERCOM.2003.1192736

[178] Zhao, Y., "Standardization of mobile phone positioning for 3G systems," *IEEE Communications Magazine*, 40(7), 2002, 108–116.

[179] Zhu, J., and Durgin, G.D., "Indoor/outdoor location of cellular handsets based on received signal strength," *Electronics Letters*, 41(1), 2005, pp. 24–26. doi:10.1049/el:20056605

# Author Biography

**Anthony LaMarca** is the associate director of Intel Research Seattle. His research interests include location technologies, ubiquitous computing, distributed systems, and human-centered design. His former project, Place Lab, enabled wide-scale device positioning using radio beacons. He is currently leading the Everyday Sensing and Perception project with the goal of building high-coverage, high-accuracy techniques for inferring common contexts. He has a BS in computer science from the University of California at Berkeley and an MS and PhD in computer science from the University of Washington. He can be contacted at anthony.lamarca@intel.com.

**Eyal de Lara** is an associate professor in the Department of Computer Science at the University of Toronto. His research interest lie in the area of mobile and pervasive computing, and in the past he has led projects in the fields of content customization for mobile devices, accurate localization in indoor environments, and secure spontaneous communication. He has a BS in computer science from the Instituto Technologico y de Estudios Superiores de Monterrey (ITESM), Mexico, and an MS and PhD in electrical and computer engineering from Rice University. He can be contacted at delara@cs.toronto.edu.

Printed in the United States
by Baker & Taylor Publisher Services